T0087273

People to Know

Barbara McClintock

Nobel Prize Geneticist

Edith Hope Fine

Enslow Publishers, Inc.
44 Fadem Road
Box 699
Springfield, NJ 07081
USA

PO Box 38
Aldershot
Hants GU12 6BP
UK

For Margaret L. A. MacVicar, friend.
Like Barbara McClintock, she is remembered
for her high intelligence, persistence, and wit.

Library of Congress Cataloging-in-Publication Data

Fine, Edith Hope.
Barbara McClintock, Nobel Prize geneticist / Edith Hope Fine.
p. cm.
Includes bibliographical references and index.
Summary: Presents the life and career of the geneticist who spent many years studying the cells of maize and in 1983 was awarded the Nobel Prize in Physiology or Medicine.
ISBN 0-89490-983-5
1. McClintock, Barbara, 1902—1992. Juvenile literature. 2. Women geneticists—United States—Biography—Juvenile literature. 3. Corn—Genetics. [1. McClintock, Barbara, 1902-1992. 2. Women geneticists. 3. Geneticists. 4. Nobel Prizes—Biography. 5. Women—Biography.] I. Title.
QH437.5.M38F54 1998
576.5'092—dc21
[B] 97-43754
CIP
AC

Printed in the United States of America

10 9 8 7 6 5 4 3 2 1

Illustration Credits:
Courtesy of Cold Spring Harbor Laboratory Archives, photo from Marcus Rhodes, p. 37; Courtesy of Cold Spring Harbor Laboratory Archives, photo by Margot Bennett, p. 6; Courtesy of Cold Spring Harbor Laboratory Archives, photo by Ross Meurer, p. 84; Courtesy of Cold Spring Harbor Laboratory Archives, photo by Sue Zehl, p. 107; Courtesy of Cold Spring Harbor Laboratory Archives, pp. 75, 87; Courtesy of Joseph G. Gall, Carnegie Institution, p. 51; Courtesy of Marjorie Bhavnani, p. 18; Courtesy of Nigel Holmes, p. 67; Courtesy of the American Philosophical Society, pp. 23, 42, 62, 93, 101; Jacob Rosenstein, p. 44; Permission of Marjorie Bhavnani, courtesy of Cold Spring Harbor Laboratory Archives, p. 15, 25; Photo by the author, p. 57; Reproduced by Greg Fine with permission from *Science*, vol. 69, issue 1798, Copyright 1929, American Association for the Advancement of Science, p. 39; Svenskt Pressfoto, Stockholm, courtesy of Cold Spring Harbor Laboratory Archives, p. 96.

Cover Illustration: Courtesy of Cold Spring Harbor Laboratory Archives

Acknowledgments

Many thanks to those who, with unfailing kindness, provided help and resources for this project: Thomas R. Broker, Ph.D., and Louise T. Chow, Ph.D., professors of biochemistry and molecular genetics, University of Alabama at Birmingham; Clare Bunce, archivist, Susan Cooper, director of institutional advancement, Trudeau Institute; and Rob Martienssen, Ph.D., Cold Spring Harbor Laboratory, Long Island, New York; Elizabeth Carroll-Horrocks, archivist, American Philosophical Society, Philadelphia; Nathaniel C. Comfort, Ph.D., George Washington University; Pat Craig, Joseph Gall, and Maxine Singer, Ph.D., Carnegie Institution of Washington; Nina Fedoroff, Ph.D., Pennsylvania State University; Jerry Feitelson, Ph.D., Mycogen Corporation; Patricia Gossel, curator, Smithsonian Institution; Mary-Lou Pardue, Ph.D., Massachusetts Institute of Technology; Ronald L. Phillips, Ph.D., Regents' Professor, University of Minnesota and chief scientist, U.S. Department of Agriculture; and Anna Marie Skalka, Ph.D., director, Institute for Cancer Research and senior vice-president, Fox Chase Cancer Center, Philadelphia.

Thanks, too, to Marjorie M. Bhavnani, Alison Cobb, Harriet Creighton, Ph.D., Pat Dorff, Anne D. Ewing, Ph.D., Janelle and Michael Fine, Nigel Holmes, Rodney le Brecque, Joan Pesek, Marc Sharp, Ph.D., and Evelyn Witkin, Ph.D. A special tip of the hat to Lee Kass, Ph.D., Cornell University, Ithaca, New York; and my writing group.

Contents

1. In the Spotlight 7

2. An Independent Child 12

3. Drawn to Science 21

4. Genetics—Monastery Garden to Cornfield 28

5. Two Momentous Discoveries 35

6. A Difficult Decade 46

7. A Home at Cold Spring Harbor 56

8. Convincing the World 65

9. In the Lab and on the Trail of Corn 73

10. Awash in Honors 81

11. Winning the Nobel Prize 90

12. Passing the Baton 95

13. A Scientific and Personal Legacy 105

Chronology 109

Chapter Notes 113

Glossary 122

Further Reading 125

Index 126

Barbara McClintock

1

In the Spotlight

Word spread fast. A press conference had been called on Long Island, east of New York City. Scientists and staff from Cold Spring Harbor Laboratory hurried to Bush Lecture Hall.

Reporters from newspapers, magazines, and radio and television stations crowded in. Microphones were on. Tapes rolled. Cameras flashed.

The spotlight was on eighty-one-year-old Barbara McClintock, a scientist who still worked daily in her laboratory. Research was the center of her life. She sat on a stool, speaking in a whisper.

"I never even thought of getting any award at this stage from the Nobel Committee," McClintock was saying.[1]

Asked how she had found out about the Nobel, she said, "I heard it on the radio this morning." Laughter crinkled her face.

"No one called you?"

"No, I don't have a phone at home, I haven't for years. When I go home, I don't want that phone to ring. . . . I want to be free."

"What will you do with the prize money?" someone called out.

McClintock, a slight woman with a short, plain haircut and a wry sense of humor, answered, "I don't even know what the award brings in."

"It's $190,000," she was told.

McClintock laughed. "Oh, it *is*? I didn't know. I'll just have to get to one side and think about this."

The Nobel Committee must have been surprised on October 10, 1983, to discover that Barbara McClintock had no telephone. For her, the lack of a phone was typical. With no need for a big house, fancy clothes, or much money, she lived a no-frills life. A light microscope was her research tool. This brilliant scientist could get along with very little.

McClintock's work centered on maize, the multicolored Indian corn often seen at Thanksgiving. She was a master in her field. All her life she had explored the fascinating world of genetics—the study of how traits pass from parents to offspring. At an age when most people had long since retired, she still walked down the road each day to work in her lab.

No one would have faulted McClintock for celebrating that October morning. She was the first woman ever to receive an unshared Nobel Prize in

Physiology or Medicine. The prize recognized her groundbreaking work in genetics.

For McClintock, the Nobel Prize was an interruption. Instead of celebrating, she set out on a solitary walk to think about how she would respond.

To anyone who knew her, this was pure McClintock. On a quiet walk through the woods and fields of the rural research facility, she could consider what winning the Nobel would involve. Impeccably dressed in her usual outfit of neatly pressed khakis, simple shirt, and pullover sweater, she was alone with her thoughts.

Corn had been at the heart of her work. For her, the maize was her teacher. Corn had taught *her* about the complex workings of genes—the tiny units of heredity that pass traits from one generation to the next. If she spotted a tiny set of red flowers hidden in a patch of airy white Queen Anne's lace, she recognized it. That unexpected red was the signature of her own extraordinary discovery, which was nicknamed "jumping genes" by newspaper reporters.

This flurry of attention surrounding McClintock was a long time coming. Many scientists had been slow to understand and recognize the importance of her discoveries.

Startling changes in her corn plants back in the early 1940s had convinced Barbara McClintock that some maize genes could move from place to place on the long chromosomes. She realized, too, that some genes controlled how others moved and how they acted.

Before the early 1970s, many leaders in the scientific world had not been ready for such news. Until then, most scientists had believed that all genes held exact, unchanging positions on chromosomes. McClintock's ideas about genes that could "jump" to new sites had seemed radical.

On her long walk before the press conference, McClintock knew that worldwide attention would soon be focused upon her. Never one to seek the limelight, McClintock now found it thrust upon her. She would have to face it.

Barbara McClintock also understood that this honor would bring attention to the Carnegie Institution and to the Cold Spring Harbor Laboratory. Carnegie had been steadfast in its support of her work. It had provided lab space, fields for growing corn, greenhouses, money, and stability. Cold Spring Harbor had been her research home for more than forty years.

Because news of the Nobel Prize had come on the Columbus Day holiday, McClintock's colleagues scrambled to organize. They ordered autumn flowers arranged along with Indian corn, answered ringing phones, wrote biographical sketches, and prepared for the impromptu press conference.

Now, in her modest way, Barbara McClintock was both earnest and bantering with reporters. She worked to translate her complex discoveries into terms the public could understand.

At the end of the press conference, she was asked how she felt about receiving the Nobel Prize. She was surprised at winning, she said, but all along she had believed in what she had discovered. The corn plant had told her very strikingly what was going on in its chromosomes. She had also known, she added, "that sooner or later it [her discovery] would be found out in other organisms. But it took a lot of time before that happened."

A reporter tossed out a last question. "What do you think of big to-dos like this? With all the attention that's being heaped on you?"

To the laughter and applause of the audience, Barbara McClintock replied, "You put up with it."

2

An Independent Child

At the turn of the twentieth century, when Barbara McClintock was born, great change was in the air.

Horses shied at the sound of automobiles, the new "horseless carriages." The Wright brothers were ushering in the age of flight. Some homes had telephones and electric lights. Silent movies thrilled audiences, and baseball was the national pastime.

A new science, genetics, came into being around that time, too. Barbara McClintock's work would be at the heart of it.

Barbara's parents, Sara Handy and Thomas McClintock, had married in 1898, despite objections from Sara's minister father, Benjamin Handy. He did not like any of the young men who courted his daughter, especially an upstart like Thomas McClintock. Thomas's parents were recent immigrants from the British Isles. The ancestors of Benjamin Handy and his wife, on the other hand, had come to this country on the *Mayflower*.[1]

Sara Handy was a lively young woman, an artist and a pianist. What did she see in Thomas McClintock? How, fretted the Reverend Handy, could Thomas McClintock possibly support a family? He was still studying to become a medical doctor.

Soon, Sara and Thomas McClintock began their family. Their first child was Marjorie, born in October 1898. Her sister Mignon arrived in 1900 as Thomas was struggling to set up his medical practice. As Benjamin Handy had predicted, the young couple faced serious financial problems.

Hoping for a boy, the McClintocks had a third child, but it was another girl. Barbara McClintock was born on June 16, 1902, in Hartford, Connecticut. As an infant, Barbara was content to entertain herself. Seldom fussy, she lay happily on a blanket or stared at a simple toy for long periods of time.[2] When she was about two, the longed-for boy, called Tom, arrived.

Sara Handy McClintock's life changed drastically when she became a wife and mother. Her well-to-do background had not prepared her for pinching pennies while raising four small children, running a household, and working. Dr. Thomas McClintock worked long hours, but money was scarce. Sara McClintock used her inherited money to pay bills left from her husband's education. To supplement his income, she began teaching piano.

Barbara and her mother were never close. Before Barbara was born, the McClintocks had chosen a name for their third child: Benjamin. Barbara knew they were disappointed that she was a girl.[3] From the start there was tension between mother and daughter. McClintock later said, "I was an odd member" of the family. Of her mother, McClintock said, "I didn't get approval, but I didn't get harsh treatment from her."[4]

Whatever the reasons for their differences, from the time Barbara was about three years old, she was sent on long visits to an aunt and uncle's home in Massachusetts. Barbara blossomed during her extended stays with Aunt Carrie—her father's half sister—and Uncle William King. Barbara was especially fond of her uncle, a fish dealer. She later described him as "a great big man with a great big voice" who would call out to the people in homes along the country roads to find out who needed fish that day.[5] Barbara loved riding along with him on his horse-drawn wagon.

Uncle William enjoyed tinkering with machinery. He bought one of the newfangled motor trucks to make his deliveries. From watching him repair it when it broke down, Barbara developed an early interest in how things work.

Her stays in Massachusetts were filled with good memories. "I enjoyed myself immensely," she remembered, adding that she was "absolutely not" homesick.[6]

Barbara McClintock, about five years old.

Time away from home did little to make things easier between the young girl and her mother. Barbara was not an affectionate child. She did not want to be hugged. Nor did she seek her mother's approval.

When she was five, Barbara's father bought her a set of small tools. Instead of being pleased with the gift, she was disappointed. The tools were toys, designed for little hands. She longed for real tools, ones that worked. Besides tools, Thomas McClintock gave her boxing gloves and other games and toys more typically given to boys.

In many ways, Barbara McClintock's childhood was a rehearsal for her life as a scientist. She could amuse herself for hours. The tiniest details caught her interest. When she read books, she was completely absorbed. On walks, she noticed plants and animals along the way. Sometimes she just sat and thought.

The fact that she was alone so much concerned her parents. For Barbara, there was nothing unusual about it. She was content on her own. Ideas soared in her mind. She would carry this curiosity and wonder into adulthood.

When she was six years old, Barbara's long visits to Massachusetts ended. The McClintocks moved from Connecticut to Flatbush in Brooklyn, New York. Unlike the bustling urban Brooklyn of today, in 1908 there were fields and open spaces. Barbara often took long walks with Mutty, the family dog.

Although quiet, Barbara did not lack confidence. Once on a family trip into Manhattan, she learned how tall the Statue of Liberty was (152 feet). "That's no problem," she said. "I can shinny up!"[7]

In the early 1900s, many parents treated girls and boys differently. With their frilly dresses, girls were expected to stay clean and avoid rough-and-tumble sports. Barbara had no interest in dolls or tea parties. She also had no close girlfriends.

Dr. and Mrs. McClintock, however, encouraged their children to pursue their own interests. Later Barbara said that her father "should have been a pediatrician because he was very perceptive with children."[8]

For the McClintocks, school was an important part of growing up, but so were nature, fresh air, and exercise. The McClintock children enjoyed a freedom unknown today. If they did not want to go to school, they did not have to go. Barbara later told colleagues that she regularly played hooky. When Barbara objected to one teacher she strongly disliked, her parents let her stay out of school for a semester.

Dr. McClintock also felt that his children should not have homework. If they worked hard during their six hours of school, they should be able to spend their free time as they wished. He wanted them to develop their own special talents.[9]

When Barbara eagerly started ice-skating, her mother and father gave her a pair of fine skates. While her fellow students studied, Barbara spent hours at Prospect Park near her house perfecting her skating.

For a time, Barbara enjoyed piano lessons. Her mother taught her at first, then Barbara had another teacher. Not surprisingly, Barbara took this piano

Barbara attended P.S. 139 in Brooklyn, New York. She is seated in the front row, at the right. Her name is written on the chalkboard as one of fourteen first graders to make the honor roll.

study very seriously. She was so highly focused that the adults stopped the lessons, fearing she was too intense about them.

Barbara was happy alone, but she was not without playmates. She and her siblings loved to play hard in the neighborhood after school. One day it was volleyball, another baseball. Being active, Barbara felt that skirts got in her way. A dressmaker sewed her a pair of bloomers—loose trousers gathered at the knee. Now Barbara was free to climb trees, run, and play sports with the boys in the neighborhood.

One day, children from her neighborhood were going to play against a team from another block. Barbara expected to join in, but the boys from her block would not let a girl play. The other team, short a player, welcomed Barbara, and that team won the game. "We beat our team thoroughly," McClintock remembered.[10]

McClintock's parents supported her, even though her ways were often different from those of other children. Once a neighbor gave Barbara a scolding. Why didn't she dress and act the way girls were supposed to act? Barbara marched straight home and told her mother. Mrs. McClintock called the woman and warned, "Don't ever do that again!"[11]

Barbara knew she was different: "I was a girl doing the kinds of things that girls were not supposed to do."[12] Looking back at her early years, McClintock once said, "I was the oddball."[13]

As a child, Barbara was free to be a tomboy. Mrs. McClintock assumed that Barbara would outgrow her boyish ways when she became a teenager. Surely

Barbara would drop her intense interest in learning, stop being so endlessly curious, and become a young lady.

Things did not turn out as her mother had hoped. In high school and college, Barbara would stay very much the same. Clothes, appearance, and socializing were not very important to her. She was content to read, study, and learn.

Barbara McClintock was reaching a turning point in her life. She would soon find a subject that drew her like a magnet.

3

Drawn to Science

At Erasmus Hall High School in Brooklyn, Barbara McClintock soared. From the start, Barbara was a focused learner. She liked thinking about things. Just as she had learned ice-skating, McClintock now spent hours and days taking in new ideas.

Many adults point to a special teacher who strongly influenced them in school. For McClintock, it was a special subject. She discovered the world of science.

Science classes at Erasmus Hall captivated Barbara and nurtured her curiosity. "I would solve some of the problems in ways that weren't the answers the instructor expected," McClintock said later. Finding the right answer to a difficult question posed by a teacher was "just pure joy," she added.[1]

In 1917 the United States entered World War I, which had begun three years earlier in Europe. Life in the McClintock family changed drastically when Dr. Thomas McClintock went overseas as a surgeon.

With her husband away, Mrs. McClintock was responsible. Years later, Barbara remembered how much her mother loved to be in charge.[2] Sara McClintock had strong ideas about her daughters and their education.

Barbara's older sisters, Marjorie and Mignon, were both fine students. They got excellent grades. Vassar College even awarded a scholarship to Marjorie, but Mrs. McClintock wanted none of it. She discouraged her older daughters from attending college.

The two older McClintock girls did what their mother and society expected of them. Marjorie became a musician. She played the harp and piano professionally before her marriage. Mignon was an actress until she married.

Barbara graduated from Erasmus Hall High School in January of 1919, when she was only sixteen. Now it was her turn to make a decision about college. Barbara was intent on pursuing her education. Her mother was adamantly opposed.

Sara McClintock wanted only the best for her daughter, but their wishes clashed. Mrs. McClintock imagined Barbara single and childless, a terrible fate in the mother's view. Her greatest fear was that her daughter would become a college professor, "a strange

Mrs. McClintock plays the piano for her four children. From left are Mignon—holding Mutty, the family dog—Tom, Barbara, and Marjorie.

person," explained Barbara, "a person that didn't belong to society."[3]

The two were ever at odds. Mrs. McClintock may have underestimated her daughter's determination to go to college. With her father abroad, Barbara didn't have his support. She was forced for the time being to put off attending college. Yet, unlike her sisters, she was absolutely set on her goal: She would get an education even if she had to teach herself.

Barbara took a job as an interviewer in an

employment agency. After work, she spent hours at the library in a course of study she designed for herself.

Barbara knew herself better than her mother did. She recognized that she was strong and unwavering. Still, she sensed how hard it was in those days to be a girl who was both intelligent and independent. She knew that the ways she thought and behaved were not, as she said later, "standard conduct, . . . but I would take the consequences for the sake of an activity that I knew would give me great pleasure."[4]

When Dr. Thomas McClintock returned home from the war, the whole picture changed. He strongly supported Barbara's wish to attend college. In September 1919, seventeen-year-old Barbara was off by train to register at Cornell University in the rolling hills of Ithaca, in upstate New York. Unlike many universities of the time, Cornell actively supported education for young women.[5]

Barbara got off to a very strong start at Cornell. At last she had found a place where she could be accepted for her brains and her independence. In her first two years at Cornell, Barbara was very social, dating and having many friends. She also became president of the women's freshman class. She played tenor banjo with a jazz group but gave this up when the late nights began to get in the way of her studies.

The years following World War I were optimistic ones in the United States. Women got the right to vote in 1920. In the "Roaring Twenties," the stock market boomed. Crowds gathered to see actress Mary Pickford, comedian Will Rogers, jazz musicians, and home-run slugger Babe Ruth.

Independent and eager to learn, Barbara McClintock settled easily into life at Cornell University.

With the new dance craze, the Charleston, young women in big cities were wearing both their skirts and their hair shorter. Barbara McClintock did not care a fig for fashion or for fads. She did care about being practical. Long hair was hot and tangled easily. Barbara had her hair bobbed, and her short hair caused quite a stir on the campus the next day.

Many of Barbara's friends at Cornell were Jewish. Invited to join a sorority, she learned that her Jewish friends were not welcome as members. Although not Jewish herself, she said later, "I thought about it for a while, and broke my pledge. . . . I just couldn't stand that kind of discrimination. It was so shocking that I never really got over it."[6]

Highly intelligent, Barbara tackled her courses eagerly. Sometimes she was so focused that she lost track of ordinary details. Years later, she recalled an incident during a geology exam. She felt well prepared and had her blue test booklet ready. "I knew there was nothing I couldn't answer," she said. "I just started writing right away and I enjoyed it. I finished very fast." Then Barbara realized she hadn't written her name on the cover of the blue book. That was a problem. Her mind was a blank. She couldn't think of her name. Embarrassed, she didn't dare ask anyone near her. "They would think me cuckoo."[7]

She sat quietly. About twenty minutes later, her name finally came to her. Her fierce concentration had wiped everything but the test topics from her mind. Such absolute focus would serve her well as a scientist.

A key event took place in her junior year.

McClintock took a course in genetics, the study of inherited characteristics. She absorbed the material, learning how plant and animal traits are passed from one generation to the next. That class would determine the course of her life.

Years later, the Nobel Prize Committee asked Barbara McClintock to tell about her life. Instead of beginning with her family or her childhood, McClintock started her life story with that Cornell genetics class.

McClintock was twenty-one years old when she graduated in 1923 with a degree in botany, the study of plants. She had also taken a special Saturday class in the study of the structure of the cell (*cytology*) with Professor Lester Sharp, and she knew for certain that she wanted to continue to explore genetics. She completed her master's degree in 1925 and her Ph.D. degree in 1927. Both were in botany.

Staying on at Cornell as an instructor and researcher, McClintock was caught up in the excitement of seeing the wonders of a new field unfold. Fascinated by genetics, she was ready to delve into the puzzles of gene behavior.

4

Genetics—Monastery Garden to Cornfield

Genetics is the study of heredity: how living things pass on traits from generation to generation. It is a twentieth-century science, very new compared with such centuries-old fields as astronomy or physics.

All living things—bacteria, yeast, night crawlers, tiger lilies, redwoods, blue whales, humans—inherit traits from their parents. The genes that determine hereditary traits are tiny units carried on chromosomes inside the nucleus of a cell. A genome is a plant or animal's complete set of chromosomes and their genes. (The word *genome* is a mix of the words "gene" and "chromosome.")

"There are really three main figures in the history

of genetics—the three M's: Mendel, Morgan, and McClintock," said Nobel medalist James Watson in 1991. "Gregor Mendel and Thomas Hunt Morgan showed us how regular the genome is, and Barbara McClintock showed us how irregular it is."[1]

To do research in genetics, scientists must choose a plant or animal that suits their experiments. Gregor Mendel studied pea plants. Thomas Hunt Morgan worked with fruit flies. Barbara McClintock focused on maize. These three scientists laid the foundation for modern genetics.

Though they lived and worked in different centuries, Barbara McClintock had much in common with Gregor Mendel, an Austrian monk. Deeply curious, both researchers delighted in nature and its infinite possibilities. Both worked by themselves. They were persistent, methodically collecting information and keeping detailed records. Both were interested in meteorology, the study of weather. In both cases, their work was overlooked for a long period of time.

Gregor Johann Mendel (1822–1884) is called the father of genetics. In his garden at the monastery, he noticed changes in pea plants from generation to generation. Some, for instance, grew tall, some short. Some bore white flowers, others purple. Some yielded green peas, others yellow peas. Some peas were wrinkled, some smooth.

Mendel was not the first to wonder why such changes took place. His genius lay in tracing and tracking certain changes over time. He broke new ground as the first person to predict patterns of inherited traits.

The word *gene* had not been invented in Mendel's time. He used the word *factors*. He learned there was something in the cell, some "factor," that determines which traits a plant or animal will inherit from its parents.

In 1856, Mendel began to study pea plants. He crossbred different varieties to see what would happen in the next generation. For example, he would take pollen from red-flowered plants to fertilize white-flowered plants. What began as a hobby became his passion. Over the next eight years, he studied about twelve thousand plants.

Mendel learned many things from his long study. One was that plants and animals inherit two forms of the gene for each trait. One form comes from each parent. These two forms (called *alleles*) can be identical or different. If different, one is usually *dominant*—that is, its effect is seen in the offspring of a plant or animal. The other form is *recessive*—its effect is hidden in the offspring but may reappear one or several generations later.

Likewise, the only time a plant will show a recessive trait is if it receives two recessive alleles for that trait. Otherwise, the dominant trait appears.

In pea plants, the pea itself is the seed. Mendel understood that his plants could have varieties in seed color and texture. For example, the coat of a pea seed could be yellow or green, round or wrinkled.

He also figured out that these traits are passed on independently: A yellow seed can be round or wrinkled, and a green seed can be round or wrinkled. Mendel realized that color and texture are inherited separately.

Dominant forms in peas are *round* seed coats (*R*) and *yellow* seed color (*Y*). Recessive forms are *wrinkled* seed coats (*r*) and *green* seed color (*y*). In one important experiment, Mendel decided to crossbreed plants that had two distinct traits. One set of plants had the dominant *round* seed coats and *yellow* seed color (*RRYY*). He crossbred those plants with others that had the recessive *wrinkled* seed coats and *green* seed color (*rryy*). The first generation's pea seeds (*RrYy*) were all round and yellow.

In the second generation, many of the pea seeds were the dominant round and yellow, and a few were the recessive wrinkled and green. Some others, however, didn't look like either parent. The traits of these other peas were a mix of dominant and recessive traits—some were wrinkled and yellow, some were round and green (see diagram, page 32).

Mendel now had clues about plant inheritance. For round, green peas to grow, the pairs of factors from the parent plants must have been rearranged independently. That allowed a round factor (*R*, dominant) and a green factor (*y*, recessive) from one cell to join a second green factor from the other cell to produce round, green peas.

Gregor Mendel worked out many interesting principles of heredity, but he wrote only two papers, and these were published in a journal that was not widely read. For many years his remarkable work went unnoticed. The careful records of his experiments

Mendel's Experiment

Mendel crossed a plant with *round* (RR) and *yellow* (YY) peas with a plant that had *wrinkled* (rr) and *green* (yy) peas. The first generation (RrYy) were all round, yellow peas. In the second generation, below, nearly half the peas showed recessive traits—green or wrinkled.

Sperm cells from *RrYy* parent (four kinds)

Egg cells from *RrYy* parent (four kinds)

	RY	Ry	rY	ry
RY	RRYY	RRYy	RrYY	RrYy
Ry	RRYy	RRyy	RrYy	Rryy
rY	RrYY	RrYy	rrYY	rrYy
ry	RrYy	Rryy	rrYy	rryy

Capital letters show dominant traits. Lowercase letters show recessive traits.

R = ROUND   *r = wrinkled (not round)* 

Y = YELLOW   *y = green (not yellow)*

were destroyed around the time of his death, so his work could not be applied to other studies of plants being done. Also, Mendel's ideas were ahead of his time—scientists were not yet ready to understand them.

In 1900, two years before Barbara McClintock was born, Mendel's two published papers were rediscovered. Other scientists were coming up with the same results as Mendel in their own experiments.

Also, microscopes had improved. Using better instruments, scientists could see parts of cells never before viewed. These scientists, known as cytologists, study how cells are formed, what their structure is, and how they work. (*Cyto-* means "cell" and *-ology* means "study of.") Now individual cells of plants and animals could be seen, as well as stringlike structures inside a cell's nucleus. They looked like snips of thread and were called *chromosomes*. (*Chromo-* means "color," and *-some* means "body.") Chromosomes took dye well, making them easier to view through microscopes.

A flurry of studies began on chromosomes. Now scientists made the connection between Mendel's ideas and the chromosomes they could see microscopically. The words *gene* and *genetics* came into use, and a new field of science was born.

Thomas Hunt Morgan (1866–1945), the second giant of genetics, studied an animal instead of a plant. For his experiments, he used tiny fruit flies, *Drosophila*.

Fruit flies live only ten to fourteen days. With a new generation born every two weeks, Morgan could study changes in many generations of fruit flies in a short period of time. *Drosophila* cells have just four different chromosomes.

From early experiments, Thomas Hunt Morgan, his wife, Lillian Morgan, and their students were able to confirm that genes are located on chromosomes in cells. They discovered that the changes in traits noted by Mendel were passed on through genes. It was also Morgan's group that found that genes have orderly positions on chromosomes. For this work, Thomas Hunt Morgan won the Nobel Prize in Physiology or Medicine in 1933. He shared the prize money with his students.

When Barbara McClintock, the third great geneticist, came on the scene, Morgan immediately recognized the brilliance of her work. He became a strong supporter of McClintock and her ideas, encouraging her at an important point early in her career. Her work would give the field of genetics extraordinary breadth and depth.

5

Two Momentous Discoveries

In the 1920s, Cornell University was famous for the study of heredity. Its department of plant breeding in the College of Agriculture focused on the genetics of maize, the corn with colored kernels. While scientists understood the basics of maize genetics, they were just beginning to study maize chromosomes.

Barbara McClintock and the botany and plant-breeding students were crammed into an old laboratory in the botany building on the Cornell campus. Working so closely together had advantages. If someone found something new or odd on a glass slide, the others could step right over for a look through the microscope. Discussion and quick debates filled the air. This lively exchange of ideas made every day exciting.[1]

Intense study of cell genetics took place at Cornell from 1928 to 1935. "Progress was rapid, the air electric," geneticist Marcus Rhoades remembered.[2] Those years have been called "The Golden Age of Maize Genetics."

Wrapped up in her explorations, Barbara McClintock worked long hours. Fortunately, her friend Esther Parker, a physician, made a place for McClintock in her home. Dr. Parker's mother took care of cleaning, meals, and laundry. McClintock was free to concentrate on her studies, thanks to these two kind, supportive women.

Barbara McClintock became recognized as the leader of a group that included Marcus Rhoades, George Beadle, and Charles Burnham. Harriet Creighton, who came to Cornell from Wellesley College, joined them. It was, McClintock said later, "a piece of fine luck [that] a very fine group of graduate students came to Cornell at the same time."[3] From the start, those bright young people worked as a team.

Unlike Mendel and Morgan, who had tracked changes in outward physical appearance, McClintock felt that much could now be learned by observing chromosome changes in maize directly under the microscope. And so her life's work began.

At that time, scientists knew that maize had ten chromosomes, but no one had yet found their distinguishing characteristics. Through a microscope, the chromosomes looked similar, like pieces of short,

McClintock with her maize group colleagues at Cornell in 1929. Standing, from left, are Charles Burnham, Marcus Rhoades, and Rollins Emerson. Kneeling with the dog is George Beadle, whom McClintock called "Beets."

curvy spaghetti. While working on a project of her own, McClintock found a way to tell corn's ten chromosomes apart.

Until then, scientists had used thin slices of cells from the root tips of maize to study cell structure. Unfortunately the long, thin chromosomes were often cut apart with that technique. John Belling had recently found a way to use dye and to "squash" the plant material instead.

Already a master of the simple light microscope, McClintock experimented, improving Belling's method. Besides using red dye, she heated the slide, then destained it (that is, removed the stain) with acid. This increased the contrast between the parts of the chromosomes that took the dye well and those that did not. With her endless patience and drive for perfection, she could prepare slides that were extraordinarily clear and beautiful.

McClintock set to work to find out more about maize's ten chromosomes. She decided to work with cells from the pollen instead of the root. In the cornfield, McClintock found a young plant whose tassel was still inside, wrapped in a whorl of leaves. The tassel would bear the pollen. At that stage of growth, cells were dividing to create new plants, the perfect time for her to study them.

Barbara McClintock cut a slit down the stalk near the top. With great care, she removed some tiny pollen cases. Each one, when mature, would hold as many as two thousand grains of yellow pollen. She became so skilled at this delicate surgery that she could close the slit, rewrap the stalk with brown paper tape, and the plant would continue to grow.

Back in the lab, she made a "smear," also called a "squash." With a sharp dissecting needle, she opened a pollen case on a glass slide and added a drop of red dye. Then she pressed gently on the pollen and "squashed" it with a cover slip—a small square of glass. This flattened the nucleus and helped separate the chromosomes.

Since some cell parts stained darker than others,

McClintock could now see differences in the ten threadlike chromosomes. It was clear that each chromosome was unique, with its own length and patterns. Some had knobs that stained very dark.

For McClintock, these differences were markers. In an important paper that she published in 1929 in the journal *Science*, she drew a diagram showing the distinct features of each maize chromosome, something unknown until then. She arranged the chromosomes from smallest to largest.[4]

She was able to work out a method for observing differences in corn chromosomes because the part of the plant she used took color better. This technique had great impact. It also highlighted her original thinking, methodical approach, and willingness to try

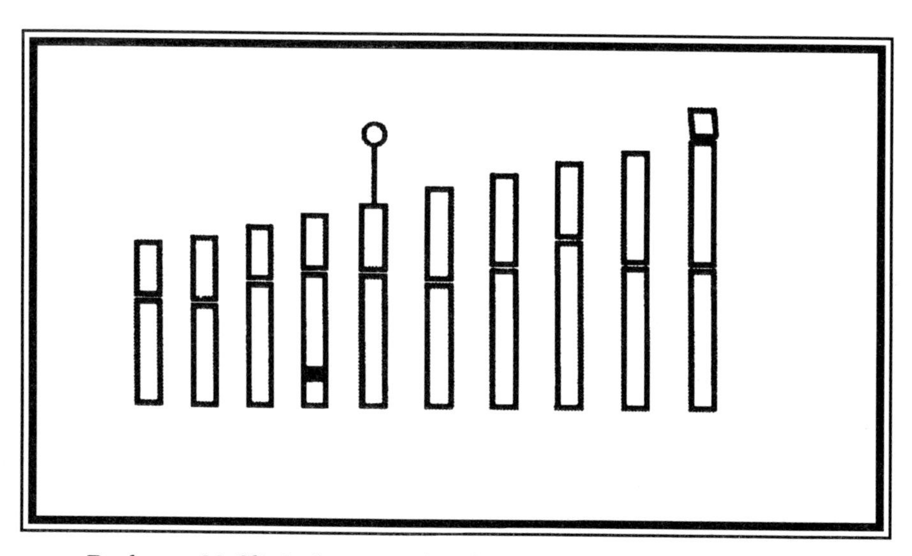

Barbara McClintock was the first to show the individual characteristics of the ten maize chromosomes. In this 1929 diagram, she arranged the chromosomes from smallest to largest.

new ideas. Now geneticists could begin to explore what jobs maize chromosomes did.

The same intense motivation that helped McClintock identify the maize chromosomes showed in both her work and her play. One June night a great rainstorm swept over Ithaca and the Cornell campus. At three in the morning, Harriet Creighton's mother called about flooding near her home. Creighton went and moved her mother to safety.

Driving back, Creighton wondered how the young corn crop was faring in the violent storm. Detouring past Cornell's garden hollow, she saw a car with its headlights aimed at the field. There was Barbara McClintock slogging about in the muddy field to save her young, foot-tall plants. She was so familiar with her maize plants that she knew their precise position in the rows. Some plants had been blown over, others torn from the ground. Wading in water, McClintock packed mud around roots. She replanted many of the numbered stalks in their original positions and saved much of her crop.

Other maize scientists had corn stock planted in that same field. Everyone stood to lose valuable research materials. McClintock was the only one who braved the storm to protect her plants.

Barbara McClintock played hard, too. On her twenty-eighth birthday, June 16, 1930, McClintock and Harriet Creighton were on the tennis court. Creighton was not quite twenty-one at the time. She thought McClintock quite old to be playing tennis so hard after working all day. Creighton's strategy was to save her energy for the shots she could return.

"Barb won by going after every shot of mine," said Creighton, "and by returning the ball where I could hardly ever reach it. Sometimes I did not even try. That brought a scathing crack from Barb about my laziness."[5]

These two incidents show McClintock's strong belief in giving her best, mentally and physically, no matter what she undertook.

Between 1929 and 1931 alone, McClintock published seven papers and co-authored two others, a large amount of published work in just three years. Each paper described exciting advances in genetics. One was the landmark paper that she and Harriet Creighton wrote in 1931.

Geneticists knew that some traits were "linked": The genes for these traits were located on the same chromosome, so the traits would almost always pass together from parent to offspring. Sometimes, though, unexpected changes took place—an offspring would inherit one trait without the other. For a long time, geneticists figured that this happened because genetic information was moving from one chromosome to another ("crossing over"), but this theory had never been proved. McClintock and Creighton set out to show that an actual exchange of material takes place between chromosomes as plants divide and reproduce.

To test this idea, they needed a strain of maize that had two markers on one chromosome, two traits that were linked and usually inherited together. McClintock had bred a strain of maize that had waxy, purple kernels. The scientists focused on its ninth chromosome. It had two markers—a knob at one end that stained well and a stretched-out tip at the other end. Perhaps, they thought, the knob related to the purple color. The stretched-out tip might relate to waxiness.

As maize geneticists, Creighton and McClintock had to be both scientists and farmers. They tilled the field, planted the seeds (corn kernels), weeded, and watered. It was hard work, often done in the heat of the day.

Corn plants are fertilized when pollen blows from the tassel (the male flower) at the top of each plant and floats to the silks (the female flower) at the ends

Maize researchers work long hours in the hot sun. Here, McClintock is labeling plants and covering the male parts (tassels that bear the pollen) and female parts (silks on the cobs) with paper bags to prevent accidental pollination.

of ears of corn. Maize researchers must control which pollen fertilizes which plants. They cannot allow pollen to float freely across the field.

To trap the pollen, Creighton and McClintock clipped small paper bags over the tassels and covered ears of corn whose silks had not yet emerged. In the field, row upon row of paper bags bobbed in the wind, like a crowd of puppets.

Then the two women hand-pollinated the silks of plants that had grown from the waxy, purple kernels (seeds). They used pollen gathered from plants whose kernels were not waxy or purple. On the daughter plants that grew from this cross-pollination, some ears had waxy, purple kernels. Others had kernels neither waxy nor purple. That was expected, given the linked traits of the parent plants.

Some ears, though, carried kernels showing only one trait—waxy but not purple, or purple but not waxy. Back in the lab, McClintock and Creighton studied the chromosomes of the new plants that showed only one trait.

Through the microscope, they saw their proof. The changes they had predicted in the ninth chromosome were there! If the knob was missing, the color purple was missing from the kernel, too. If the chromosome's stretched-out tip was missing, so was waxiness in the kernel.

Tiny sections of the chromosomes had actually switched places as the cells divided during meiosis, giving the daughter plants genes with traits different from the parent plants. McClintock and Creighton showed that genes for physical traits were carried on the chromosomes and that changes in traits were caused by an actual exchange of chromosomal material.

Proving this "cross-over" of genes, also called "recombination," was an extraordinary accomplishment. Creighton and McClintock had unlocked a basic attribute of chromosomes, demonstrating why offspring, whether plant or animal (including humans), can be different from parents. This vital contribution laid the groundwork for all future genetic research.

Crossing Over

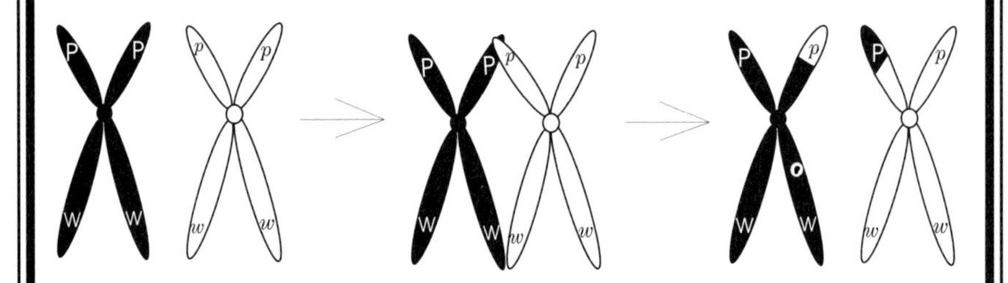

For maize, the genetic markers for the traits "purple" and "waxy" are linked—that is, both are carried on the same chromosome. Traits carried on the same chromosome are inherited together. But sometimes sections of chromosomes trade places ("crossing over"). This creates variety in the offspring.

P = purple W = waxy *p* = not purple *w* = not waxy

Always cautious, McClintock planned to confirm her findings with a second maize crop. Fortunately for Creighton and McClintock, the famed geneticist Thomas Hunt Morgan was visiting Cornell. He saw the importance of their experiment at once.

Morgan's message was clear. He advised McClintock and Creighton to publish their results immediately. Instead of waiting three months for another maize crop to grow, the two researchers quickly published a paper describing what they had found. Had they waited, German geneticist Curt Stern would have been the first to report evidence of crossing over from his studies of fruit flies.

The cross-over study by Harriet Creighton and Barbara McClintock has been called one of modern biology's greatest experiments.[6] The 1931 paper explained an aspect of chromosome behavior that had never been proved before. Many people felt the paper on crossing over and McClintock's other early papers were so outstanding that she deserved a Nobel Prize for her work.[7]

That did not happen.

6

A Difficult Decade

Two years before Creighton and McClintock's momentous 1931 paper was published, the American optimism of the 1920s had been shattered. On "Black Friday" in October 1929, the stock market crashed. High-priced stocks were suddenly worth only a few dollars. Overnight, people across the country went from wealth to ruin. This economic disaster brought on the Great Depression. Research money, always hard to come by, became even scarcer, especially for female scientists.

McClintock had a job as an instructor at Cornell. She could have kept that job. As a leader in her field, Barbara McClintock had great respect from her male colleagues. "She was something special," Marcus

Rhoades said. "I've known a lot of famous scientists, but the only one I thought really was a genius was McClintock."[1]

Teaching jobs were also available for women at women's colleges. Still, students were used to male professors. One bright graduate student once told McClintock that he could not stand being taught by women professors. "He went on and on," she remembered. "Finally I said, 'To whom are you talking?'"[2]

In the 1930s, female professors were rare at research universities. Those women who trained in science were doing research at universities either as research assistants or as wives of scientists. Despite strong contributions, these women were often considered helpers, not scientists in their own right. Barbara McClintock was not interested in teaching at a college for women or in being the wife of a scientist.

Research-driven and independent, McClintock was determined to continue her work. She wanted more than anything else to explore and solve the mysteries of maize genetics. Once, while driving her beloved Model A Ford, her mind was on several promising young scientists who had died in car accidents. She worried that if she were killed, then she would never have her answers: "This was my one overriding fear."[3]

Like many other scientists, Barbara McClintock found that research jobs often did not last long. From 1931 to 1936, she moved from one research position to another as funding ran out. At the time, good highways were few and far between. She crisscrossed the country in her Model A, many times driving on dirt farm roads that tested her resourcefulness. McClintock often had to fix the car, making repairs or mending flat tires.

Several large charitable organizations gave money to support the sciences. Unfortunately, this money most often went to support young *men* in science. Few women scientists at the time were awarded fellowships, funds for advanced studies. Barbara McClintock, however, was honored with three major fellowships.

The Rockefeller Foundation had begun giving National Research Council (NRC) fellowships in 1919. Between 1920 and 1938, eighty-six men received NRC funding in botany. Just five women botanists received NRC support in those eighteen years. Barbara McClintock was one. Her 1931 NRC fellowship supported her during a two-year period at the University of Missouri, the California Institute of Technology (Caltech), and Cornell.

McClintock drove to the University of Missouri at Columbia in the summer of 1931. She joined Lewis Stadler, a friend and maize geneticist, to study the effects of X-rays on maize chromosomes.

A few years earlier, scientists had discovered that X-rays could cause breaks and other chromosomal changes, called *mutations*. By studying mutations, scientists could learn more about genetics. Mutations occurred much more often in X-rayed materials than they did in nature. Mutations caused changes in the color and shape of kernels and leaves. Such changes could be traced to the chromosomes.

Again, McClintock made another major discovery. She had noticed unusual differences in some of the mutated plants. In microscopic slides of those plants, McClintock saw broken chromosomes, their ends frayed like worn rope. She reasoned that the randomly broken ends could fuse into a ring, because a ring chromosome would cause the mutations she had seen in her plants. Other scientists agreed.

When she finally viewed an actual ring chromosome in a microscope the next winter, she saw that she had been right. McClintock had figured out that ring chromosomes existed before ever seeing one. (She would call this fraying-healing process the breakage-fusion-bridge cycle.)

Her next job was in California. Driving her Model A west, she went to study at Caltech in Pasadena. Caltech was a men's college at the time. The school, however, made an exception for McClintock. She was the first woman ever to be accepted. She was part of Caltech's new division of biology headed by Thomas Hunt Morgan, who continued his study of fruit flies. McClintock's friends George Beadle and Charles Burnham were also at Caltech.

Geneticist James Bonner remembered her as a small, slight person. With her dry sense of humor, she smiled more than she laughed. She was quiet and worked hard.[4] Sometimes, she joined others in parties, picnics, and trips to the beach, desert, or mountains not far from the Caltech campus.

Bonner's lab was close to McClintock's. During one of their chats about genetics and life, she mentioned that she planned to stay single. She said she had met only two men worth having for husbands, both of them geneticists, but they were not available.

At Caltech, McClintock did another fine piece of detective work. She had noticed a curious structure usually seen at the end of chromosome 6 in maize. She alone was pursuing it. She called it the "organizer" when she discovered that its role was to organize the materials that would become the small, dense region (called the nucleolus) inside the nucleus. "I don't know why, but I was sure it had the key," she said.[5]

McClintock found that when the organizer was split (in this case artificially and accidentally by radiation), so there were two separate "half" organizers in the nucleus, each formed—or "organized"—an independent nucleolus. Like much of McClintock's later work, the paper she wrote in 1933 about this nucleolar organizer region (NOR) on chromosome 6 was not widely read or understood.

Not until the 1960s was the nucleolus found to be a factory, manufacturing material vital for making protein. That explained the division process McClintock had observed three decades before.

Just as her two-year NRC fellowship ran out in 1933, Thomas Hunt Morgan, Lewis Stadler, and Rollins Emerson recommended McClintock for a coveted Guggenheim Fellowship. Again, she beat the odds. Only five of the 144 scientists who had received a Guggenheim since 1925 were women. With this new fellowship, she would study in Germany. In ordinary times, a Guggenheim would have been a sure advance for her career. But times were far from normal.

In 1933, Adolf Hitler and his Nazi Party had taken control of Germany. Hitler hated Jews, people of color, and disabled persons, among others. Hitler took over the German universities, firing all Jewish professors and administrators. Germany was in turmoil.

McClintock left for Germany late in 1933, unaware of how bad things would become. Many prominent Jewish scientists, including Curt Stern, had already fled the country. McClintock had many Jewish friends and colleagues whom she respected and cared about. Appalled at the hatred and fear she found in Germany, she left abruptly four and a half months later.

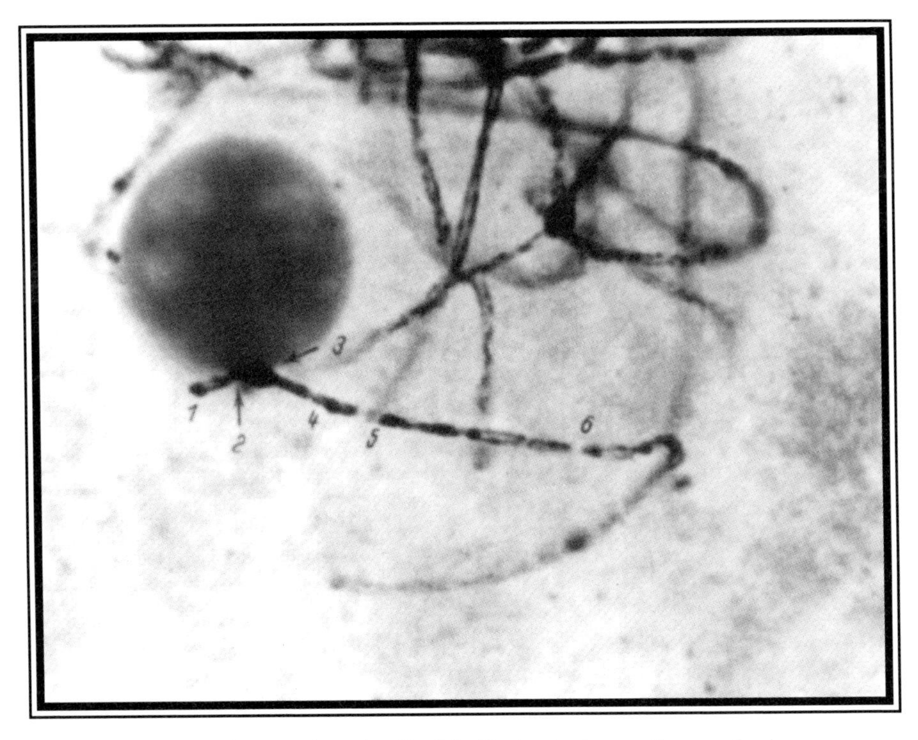

This 1934 photo taken by Dr. McClintock shows the nucleolus as a large round object on the left. The nucleolar organizer region *(NOR) is the dark bump on the chromosome numbered 3. (The maize chromosomes appear as thin lines.)*

When she returned to Cornell, her home base, she found the United States still in the grip of the Great Depression. Her fame was growing in scientific circles, but Barbara McClintock was again caught without money or a research home.

Once more, friends helped. Rollins Emerson and Thomas Hunt Morgan convinced the Rockefeller Foundation to fill the gap by providing McClintock with $1,800 for the year. The grant was renewed the following year. This was the most money McClintock had ever earned in her life.

With Rockefeller support, McClintock worked at Cornell in Emerson's lab from 1934 to 1936. She published only two papers in 1935 and none in 1936. Despite the advances she had brought to genetics, McClintock realized that staying at Cornell was not an option. There were only eight women on the science faculty. Five of them were in nutrition, which was then thought of as a woman's field.

Although she had Emerson's strong backing, McClintock could not become a permanent member of Cornell's staff. There were women at Cornell in nonfaculty positions—instructors, assistants, and graduate students. However, neither the plant breeding department nor the botany department had any women as full professors.

The same situation was true at most major U.S. universities. Few women held positions in the sciences. Those who did were seldom full professors, and rarely heads of departments.

As Barbara McClintock's third fellowship ran out, it was Lewis Stadler who found her a position. Stadler was establishing a genetics department at the University of Missouri and McClintock would be an assistant professor.

Barbara McClintock arrived at the campus in 1936, at the age of thirty-four. She was not made to feel welcome. She was not invited to faculty meetings, and she understood quickly that she was unlikely to be promoted.[6] She was there only because of Stadler's backing.

The university, located in Columbia, Missouri, was a conservative one. People were expected to fit into the small, traditional midwestern town. Those who did not remained outsiders. McClintock, with her knickers and bobbed hair, stood out. She knew her mind and spoke it, regardless of how women were "supposed" to act.

Here, she taught as well as doing research. Her class lectures were dense with information, and she expected her students to work hard and to think hard. Barbara McClintock had little patience for the inept or unprepared. One graduate student found McClintock so intimidating that when she entered the greenhouse by the front door, he fled by a back door.[7]

McClintock's ways did little to endear her to people on the campus. She had a strong personality and could be abrupt. She had run-ins with other faculty members and administrators. She had no regard for mindless conformity.

Once, McClintock forgot her key to the Agricultural School lab, so she climbed in through a window. Unfortunately, she was seen, and the story spread. Such behavior, though resourceful, was not considered fitting for a teacher and researcher. Being dignified, however, was not one of McClintock's highest aims.

She faced other prejudices, too. An engagement announcement for a "Miss Barbara McClintock" appeared in a Missouri newspaper in 1936. When the chairman of the botany department read it, he did not know it was for another young woman with the same name. Assuming it referred to the scientist he knew, he warned McClintock that she would be fired if she married.[8]

True to her style, McClintock forged ahead with her research despite the hostile atmosphere. She and Stadler continued to track changes in plants grown from X-rayed corn seeds. During her five years at Missouri, she continued to explore the complex breakage-fusion-bridge cycle.

McClintock's research advanced, and her career continued upward. She was elected vice-president of the prestigious Genetics Society of America in 1939. Despite her reputation and exceptional research, she was still thought of as stubborn and uncooperative at the University of Missouri. Her situation grew rockier.

McClintock asked the dean directly if she could ever expect to become a tenured faculty member at the University of Missouri. He said it was unlikely. He added that she would probably be fired when her mentor, Lewis Stadler, left.

The university community might have tolerated her quirks had she been a man, but she recognized that because she was both a maverick and a woman, her idiosyncrasies were not considered acceptable. McClintock found herself unhappy and out of place at the University of Missouri. She saw no future there. It was "awful, awful, awful," she said years later.[9] She knew she would never fit into the academic life on a university campus.

McClintock was at the peak of her career, yet her job was not secure. Briefly, she considered studying meteorology and becoming a weather forecaster. Leaving the university seemed to mean the end of her life's passion—digging still deeper into the study of genetics.

She took a step in 1941 that would change her life. She appealed to Marcus Rhoades, her old friend from Cornell University. He was growing his summer corn crop at Cold Spring Harbor on Long Island Sound, thirty-five miles east of New York City. Could she join him?

When he said yes, McClintock took a leave of absence from the university, then packed her Model A and headed for the East Coast, leaving Missouri behind.

7

A Home at Cold Spring Harbor

Cold Spring Harbor Laboratory faces a deep V-shaped harbor on the northern coastline of Long Island. Barbara McClintock settled into the serene setting, planted a corn crop, and tried to decide what to do next. Should she give in and return to the University of Missouri? Should she try for a position elsewhere? Should she find a way to stay at Cold Spring Harbor?

Milislav Demerec, director of the Carnegie Institution's department of genetics, based at Cold Spring Harbor, urged Carnegie to offer McClintock a position. He called her "a brilliant research worker, with keen analytical powers and the perseverance to pursue a problem through its finest ramifications."[1]

Learning of Carnegie's interest in McClintock, the University of Missouri wanted her back. Not surprisingly, after her experiences at Missouri, Barbara McClintock preferred to stay at Cold Spring Harbor. She accepted a one-year position in Demerec's department. The next year, she became a regular Carnegie staff member at Cold Spring Harbor. At last, McClintock had a place to live, grow corn, and do her research. She also had a salary.

"You couldn't mention a better job. It was really no job at all," said McClintock.[2] The Cold Spring Harbor Laboratory setting was ideal for McClintock.

This building on the grounds of Cold Spring Harbor Laboratory was known as the Animal House when McClintock worked there in the 1940s.

Buildings nestle along the base of a steep glacial hill. Many labs and offices have a breathtaking view of the harbor. The little village of Cold Spring Harbor lies across the water on the eastern shore.

McClintock blended into the campus atmosphere where scientists worked intently on varied research projects. The informal dress codes suited her sensible work outfits. Here she did not feel like an outsider or have to chafe against other people's ideas of how a woman and scientist should act. She fit in with the other researchers, whose long lab hours often matched her own.

With Carnegie support, McClintock was her own boss. Pressure to teach or publish fell away. She could work at her own pace and in her own style. Unlike, some scientists, whose work was interrupted while the United States fought in World War II from 1941 to 1945, McClintock quietly continued her research.

Her Uncle William's machinery had fascinated McClintock as a child. As a scientist, she still liked to work with machines and equipment. She cleaned and cared for the microscopes in her lab and did all her own car maintenance.

McClintock's cornfield was near her laboratory. Since she carried out lab and field research seven days a week, friends met there, rather than at her tiny sleeping quarters. Her unheated, two-room cottage was nothing more than a converted garage.

Even as she settled in, McClintock began to find herself more isolated in her research. Her focus was on the structure of the chromosome. Optical microscopes were her tools. Many scientists, though, were beginning to think the study of cells was old-fashioned. They doubted that much more could be learned from research like hers. Instead they turned to a new field, molecular biology, and studied the chemical arrangement of molecules in cells.

Electron microscopes, far more powerful than McClintock's optical microscopes, had been invented. These microscopes use focused beams of electrons to form an enlarged image. It was an exciting time for molecular biologists, but they were looking forward, not back. To many, McClintock's findings were no longer relevant. But Barbara McClintock stayed on her chosen path. For several years, she continued studying the breakage-fusion-bridge cycle, cross-breeding X-rayed stock.

• • • • •

For McClintock, 1944 and 1945 were landmark years. She made a highly successful detour into the mysteries of pink bread mold and received two high honors.

In 1944, McClintock took a ten-week break to visit George Beadle, now at Stanford University in Palo Alto, California. He had invited her to his lab to try to identify the chromosomes of a pinkish bread mold called *Neurospora*. The mold chromosomes were much tinier than her familiar corn chromosomes. Little was known of the mold or how its cells divided.

While at Stanford, McClintock wrote a letter to Milislav Demerec:

> *I felt considerable fright in tackling Neurospora chromosomes, so set myself at the task right after arriving to see just what sort of a mess I was really in. After working all day and burning much midnight oil, I feel considerably improved in my confidence. It seems to be turning out all right and I am beginning to relax and enjoy it.*[3]

All her years of cell work helped her with this unfamiliar organism. She succeeded in identifying the *Neurospora* chromosomes and learning about their cell-division process.

That same year, when Barbara McClintock was just forty-one years old, she became the third woman to be elected to the National Academy of Sciences in its eighty-one-year history. McClintock reacted not with awe or gratitude but in her own way. Because the organization had named only two other women before her, she felt a burden to succeed. If she were a man, the pressure would not be there. "It was awful," said McClintock, "because of the responsibility to women. I couldn't let them down." In a letter to a friend, she wrote, "I am not a feminist, but I am always gratified when illogical barriers are broken—for Jews, women, Negroes, etc. It helps all of us."[4]

McClintock's story could have ended there. Instead, these were just inklings of things to come. Homing in on an oddity in her maize plants, she began some of her most remarkable research.

She had been working steadily on mutations. Looking at her 1944 maize crop, McClintock noticed unusual patterns of color like those in spider plants and other garden and house plants with variegated leaves. In the corn, some yellow or pale green leaves had twin sections—one with only a few dark green stripes, and the other with many stripes. She surmised that "one daughter cell had gained something that the other daughter cell had lost."[5] Kernels, too, showed odd features—big splotches, smaller color patches, or tiny speckles instead of the usual solid colors.

Rather than being dismayed with the unexpected results, she set out to discover what had happened. What caused these peculiar new blotches on the kernels and the colorless stripes on the leaves? "I was very curious about this. Something was happening to change the action of the genes," she said.[6]

"How was it being done?" she asked herself. "There had to be an outside influence."[7] Her precise genetic mapping gave her ideas. There must be a "switch" that signaled when a gene should turn on or off during plant development. Some genes, she thought, must move about on the chromosomes. This was a revolutionary idea.

Evelyn Witkin, then a graduate student, recalls visits to Barbara McClintock's laboratory in 1944. "I was able to look over her shoulder often as her careful experiments began to tell her what the patterns of spots on her maize kernels meant and to follow her patient explanations as she increasingly understood [what was taking place]."[8]

Filled with excitement, McClintock often called Witkin in to see and talk about her increasing understanding of how some genes could change positions on the chromosomes. "How startling that was [then]!"

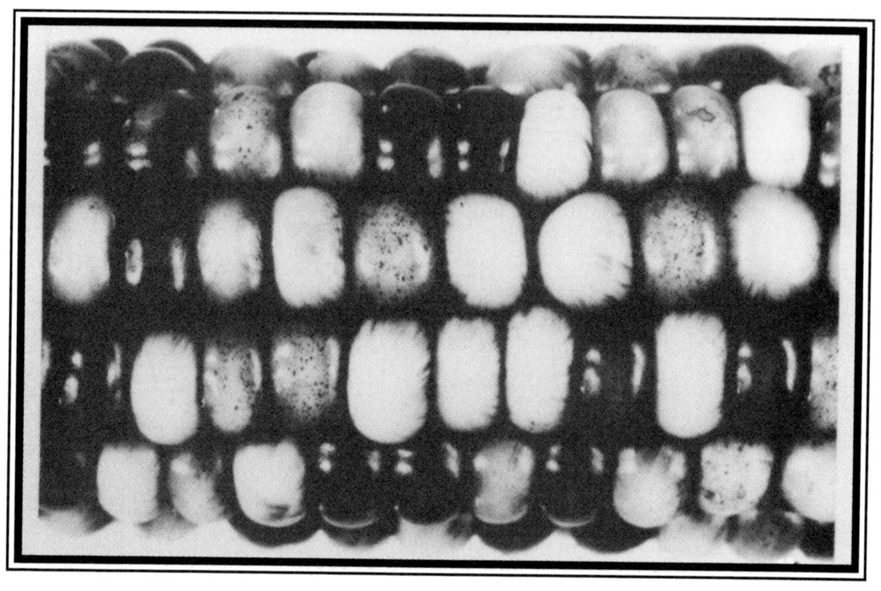

Just by looking at corn kernels, Barbara McClintock knew when in the cell cycle a mutation had occurred. A large blotch of color meant the mutation took place in the early development of the kernel. The smaller the speckles, the later the mutations took place in the kernel's development. A single speck meant a change happened in the very last stage of division.

recalled Witkin.[9] Witkin, now retired, went on to hold the Barbara McClintock Chair at Rutgers University in New Jersey.

For McClintock, her corn crops were like family. She knew the parent crops and recognized the changes that occurred in each generation. No one else could match the wealth of experience McClintock had accumulated in two decades of maize study. She could look at a slide and see things others could not.

In 1945, the Genetics Society of America elected her president. She was the first woman ever to hold that post. This important recognition was based on her work to that point. McClintock had moved genetics forward in dramatic ways.

McClintock went deeper and deeper into her work. Over the seasons, she labored on, raising crops in the summer and studying results the rest of the year. For the next six years, McClintock asked a basic question: What controls gene action? She worked with steady confidence, puzzling through massive amounts of evidence, marching toward startling conclusions. Her work was even more remarkable because the chemical makeup of genes, DNA (*deoxyribonucleic acid*), was understood by only a few scientists in the 1940s.

At this time, McClintock was spending twelve to sixteen hours each day, seven days a week, in the lab or working in her fields. When scarecrows didn't stop raccoons from feasting on her crops at night, she lugged a sleeping bag out to the cornfield. She would awaken when she heard the hungry raccoons rattling through the rows of corn and chase them off.

Further recognition came her way in 1946. Barbara McClintock became one of fewer than twenty women ever elected to the American Philosophical Society (APS), founded by Benjamin Franklin in 1743. (In 1940, only four of its 437 members were women.) Distinguished APS members had included Charles Darwin, Marie Curie, Albert Einstein, more than a dozen presidents, and a number of Nobel Prize winners.

The following year, 1947, the American Association of University Women (AAUW) was one of the first national organizations to recognize McClintock's research with a financial prize. Their $2,500 Achievement Award cited work that "yielded epoch-making results . . . with brilliant promise of still further achievement."[10]

Accepting the award in Dallas, Texas, McClintock called for greater support of young people interested in science, especially women. "On these young scientists we must concentrate much of our effort," she said, "if we wish to preserve this potential source of cultural wealth."[11] The AAUW award was especially significant because it offered opportunities for women like McClintock at a time when 95 percent of awards from other large programs were going to men.

In the late 1940s, a small group of geneticists gave Barbara McClintock a secret nickname that was both fitting and endearing. McClintock often wore blue denims while working in the cornfield. The metal buttons on her denim jacket bore the brand name "Big Mac!" Long before the McDonald's Big Mac was a familiar fast food, these colleagues dubbed her "Big Mac." This nickname didn't fit McClintock's tiny frame, but it did match her insight and intellect.

Working on her own in a small lab near the water's edge of Cold Spring Harbor, Barbara McClintock had found that not all genes stay in fixed locations on chromosomes and that some genes tell others how to act. By the late 1940s, Barbara McClintock knew she had made a discovery that would shatter commonly held views of genes.

Now all she had to do was convince the world about "jumping genes."

8

Convincing the World

What Barbara McClintock was proposing was a radical notion—that genes could move to new sites.

In the 1940s, most scientists believed that all genes had set positions on chromosomes, like pearls on a string. McClintock's wealth of evidence said otherwise. Not only could some genes "jump," but a complex system controlled their actions. She called these mobile genetic elements *transposons*. The press would later dub them "jumping genes."

As far as Barbara McClintock was concerned, the picture was clear. Even though her findings went against everything understood by geneticists, McClintock knew she was right. Another person studying these leaf and kernel changes might have thought they were a fluke. Perhaps an accident had flawed the results. Perhaps the records were not accurate.

This was not so with Barbara McClintock.

Anyone who knew McClintock also knew her reputation. Like a sleuth, she gathered clues with great care. In 1951, after six years of intense research, Barbara McClintock was ready to present her new view of genes at a major meeting at Cold Spring Harbor. She wrote a paper filled with evidence, sure that her findings could not be contested.

The scientist, now forty-nine, set about explaining what happens at the gene level to cause changes in color, texture, or pattern in maize. The plant had taught her that another gene actually *moved* to the site of a gene that caused color. She called it the *Ds*, the *dissociator*. The *Ds* told the color gene whether or not to do its job. Making this even more complex, the *Ds* could be controlled by another gene, which McClintock called the *Ac*, or *activator*.

Years later, artist Nigel Holmes used a cartoon format to help the public grasp this major discovery (see page 67). For *Time* magazine, he drew three characters—a painter, a boss, and a police officer, all standing on a curvy X-shaped chromosome. The boss (the *Ds*, or *dissociator* gene) can jump over to the painter (the color gene) and give an order. "Paint!" the boss may say, or "Don't Paint!"

The boss, however, must obey the police officer (the *Ac*, or *activator* gene). Sometimes, the officer tells the boss to let the painter do his job—color the kernel. When this order is given, the kernel that grows

This cartoon created for Time *magazine shows a gene "jumping" on a strand of a chromosome. McClintock discovered that genes could move from one place on a chromosome to another, or to a different chromosome in the same cell.*

has a solid color. Sometimes the officer tells the boss to have the painter stop, then start up again. This causes a kernel to be speckled. If the police officer does not interfere, the boss can take the painter's paint away, not letting him do his job. Then the kernel will be colorless.

McClintock knew that the notion of mobile genes went against all the accepted ideas of the time. But she expected her fellow scientists to share her excitement, ask questions, provide support. None of that happened. Like her work, the report was long, complex, even tedious in detail. Harriet Creighton recalled, "It fell like a lead balloon."[1]

Marcus Rhoades recognized McClintock as a great scientist and considered her a cherished friend. Always her champion, he never doubted her findings. "If Marcus Rhoades hadn't told me it was true," said another scientist, "I wouldn't believe a word."[2]

Some geneticists, especially those working with fruit flies and maize, kept an open mind. They were intrigued by the possibilities that McClintock's findings opened up.

Alfred H. Sturtevant, a well-known Caltech geneticist, was asked his reaction to McClintock's paper. He said, "I didn't understand one word she said, but if she says it is so, it must be so!"[3]

Others were more skeptical. McClintock later reported that one scientist told her, "I understand you're doing something that's very strange. I don't want to hear a word about it."[4]

Geneticist Oliver E. Nelson, Jr., recalled discussing McClintock's 1951 paper with a maize geneticist. The man's face reddened when talk turned to "jumping genes." His voice got more and more shrill until "he was literally jumping in his indignation . . . , [saying] he had never heard anything as ridiculous."[5]

Reactions such as these "were fun," McClintock reported. She never doubted her discovery. "There was no question in my mind," she said later, "that the information I had received [from the plant] was so strong, there was no way of drawing any other conclusion."[6]

Many scientists failed to see the importance of McClintock's 1951 report. Some simply lacked the background and expertise to understand it. The complex paper was hard to follow, a problem McClintock had experienced before. Other people felt that while McClintock's ideas were interesting, they applied only to corn.

People close to her knew that at times McClintock sometimes felt discouraged and annoyed that her paper had not been understood. Despite this, she simply moved forward. Secure in her belief in her findings, the lack of response from other scientists meant little to her. "Nobody can hurt you," she said. "You can't be turned off."[7] Her work was so fulfilling that she would have done it for free. "I could hardly wait to get up in the morning," she said.[8]

As a part of the lab community, she was aware of other researchers and their projects. Always willing to talk over scientific problems with others, McClintock listened fully and gave thoughtful input. In staff meetings and seminars, she made stimulating comments. She offered insights, suggestions, and helpful criticisms when asked. She is remembered for her clever remarks, dry wit, and hearty laugh.

She relished the natural world, just as she had as a child. "I love the springs, summers, and falls for all they can entertain you with," she said.[9]

Barbara McClintock asked chemist Waclaw

Szybalski, new to the staff in 1950, to join her one day on her daily walk. "We walked for many miles with Barbara showing me many features of the Long Island Sound shore, lakes, and wood," said Szybalski. "During the following months and years, we must have covered hundreds of miles, taking several-mile walks two or three times every week."[10]

McClintock published an article in the scientific journal *Genetics* in 1953, adding details on a second controlling system she had found. She received only two requests for copies of her article.

Because her work was having little impact, McClintock decided she was not going to bother much with getting her research results published in the major scientific journals or to look for their approval. This was an unusual decision for a scientist.

From then on, most of McClintock's reports were summaries in the Carnegie Institution of Washington's annual yearbook. This publication was not widely read and few libraries subscribed to it. Instead, she kept her own detailed records. She tracked her experiments with great precision. Her files were filled with photographs and detailed notes on index cards.

Also in 1953, James Watson and Francis Crick made a major advance when they described DNA's double helix, like a tiny twisted ladder, that makes up genes. DNA, just 1 ten-millionth of an inch thick, contains the genetic code and transmits the hereditary pattern in all living things. Now that scientists knew the structure of DNA and its key role in heredity, they began to shift their focus from studying cell structure, as Barbara McClintock did, to understanding the molecular makeup and chemistry of genes.

Waclaw Szybalski suggested to McClintock that it was time to understand genetics in chemical terms. "Waclaw," McClintock replied, "you are just a chemist; you have no feelings for biology and you do not realize how complicated it is."

"Barbara," said Szybalski, "your work would become much more clear to everybody if your jumping genetic elements could be demonstrated in bacteria, especially in *E. coli.*"

"Yes, Waclaw," Barbara answered politely, "it would be very nice, because I'm afraid that very few scientists understand my work."[11] (Szybalski himself was to become a highly regarded geneticist.)

In 1956, McClintock got ready to present her work again. She took care writing the speech, knowing more about the genetic reasons for color changes, splotches, streaks, and stripes caused by mutated chromosomes in her maize crops. With more proof and more in-depth experience, she hoped for a better reception. "[It] would be surprising indeed if controlling elements were not found in other organisms, for their prevalence in maize is now well established," she said, ending her speech.[12]

The hoped-for response failed to happen. Once again, other scientists were taken aback by the length, depth, and complexity of McClintock's ideas. The spark that burned inside her failed to ignite the general scientific community.

With each year of research, it became more and more difficult for McClintock to explain her work and have it understood by people who lacked her expertise. It became harder and harder for scientists who had not spent years with this specific maize problem to take in its significance.

Still, there were some scientists who did comprehend and respect McClintock's work. Over the years, as researchers went back to read her papers, doors of understanding began to open. Reading her papers was hard work. Even today, students and scientists find it takes real effort to trace and retrace her ideas.

McClintock was respected enough to be one of just nine women scientists elected to two or more major honorary academies from 1940 to 1968. She had already been made a member of the National Academy of Sciences (1944) and the American Philosophical Society (1946). The American Academy of Arts and Science, the largest of the three, elected her a member in 1959. Only she and two other women were members of all three.

For two winters in the late 1950s, McClintock took time away from her lab to go to Central and South America. With support from the National Academy of Sciences, she trained geneticists to do cell studies of maize. While there, she learned Spanish. On her return home, she watched Spanish-language television stations to keep up her skills.

Barbara McClintock was stubborn. She kept at her research, even when her findings were largely ignored. She worked on, confident in her discoveries.

9

In the Lab and on the Trail of Corn

By 1960, Barbara McClintock's exciting ideas about "jumping genes" were still not widely known. In the early 1960s, high school and college biology students were still being taught that all genes stayed in set places on chromosomes.

In 1960, the French geneticists François Jacob and Jacques Monod published a paper that mentioned McClintock's work. Studying *E. coli*, bacteria often found in the intestines, they discovered a controlling system much like the one she had described.

Jacob and Monod's approach was different. They were using electron microscopes to study the chemistry of bacteria and had not yet found "jumping genes" in bacteria. Still, at a 1961 meeting at Cold

Spring Harbor, they credited McClintock with finding a system of gene control. (Jacob and Monod won a Nobel Prize in 1965 for their work.)

McClintock was excited to have other geneticists confirming aspects of her work. The eyes of the science world, however, stayed focused on Jacob and Monod's studies, not McClintock's.

During the 1960s, McClintock took some time away from her lab each year. She went on the trail of corn in Central and South America from 1963 to 1969. She worked as a consultant to the Agricultural Science Program of the Rockefeller Foundation to train geneticists.

Working with Takeo A. Kato Yamakake of Mexico and Almiro Blumenschein of Brazil, McClintock helped trace the geographic origins of maize. The Spanish she had learned on previous visits came in handy.

Knobs on maize chromosomes were the basis for this study. Seeds from many plants travel easily on their own. Dandelion puffs become airborne parachutes. Burs travel on animals' coats and cling to people's socks. Some seeds spill or pop from their cases.

Corn seeds—kernels—are different. Each kernel clings tightly to the cob, and the cob itself is snugly wrapped. Corn seeds can travel only when humans carry them along as they move from place to place. At journey's end, the corn can be planted for food.

McClintock and her colleagues studied maize samples from sites hundreds of miles apart. She and others hunted for chromosomes with the same number, size,

Barbara McClintock spent long hours analyzing the patterns of change in maize kernels. One day, when she had unlocked a particularly puzzling mystery, she was so excited that she ran from her lab to the cornfield behind the building shouting, "Eureka! Eureka!"

and locations of certain knobs. Look-alike chromosomes showed that the sample came from a certain strain of ancient maize stock. In the end, McClintock and her team could plot the distinct knob locations on geographic maps.

"The completed maps indicate that there were at least four, and probably five, separate origins of present-day maize," wrote McClintock. From these maps, "migrations from each of these centers can be traced. . . . [to] reveal the movements and associations of earlier tribes."[1] This microscopic study of corn chromosomes showed a pattern of the journeys of ancient American Indians. Later she wrote a book with Blumenschein and Kato Yamakake about this research.

After her travels, Barbara McClintock was always content to return to her lab. She was now in her sixties, a time when many scientists would be making plans for well-earned retirements. But McClintock kept steadily at her work.

Although little disturbed her daily routine, McClintock always made time for friends and colleagues. Anna Marie Skalka, now director of the Fox Chase Cancer Center in Philadelphia, was once McClintock's neighbor. Skalka remembers often looking out her living room window and seeing McClintock walking along the nearby seawall. Sometimes she would stop at the back porch for a chat with Skalka's young daughter. "It was a touching scene," said Skalka, "the diminutive scientist and the tiny little girl, in congenial but serious conversation."[2]

McClintock returned to Cornell University in 1965

as part of a new visiting professor program. Recognizing her achievements as a scholar and a scientist, Cornell appointed her one of the first five people to participate in this program. She would be both a teacher and a mentor.

James A. Perkins, president of Cornell, had written her a letter confirming the appointment. He enclosed a booklet describing each of the five participants. McClintock, who knew many people thought she was unconventional, responded: "The descriptions also made me recognize that you were forewarned of some of my eccentricities. It is good of you to forgive them."[3]

Over the next six years, McClintock traveled back to Cornell each year to give lectures. A host recalled that she "seemed more comfortable in the laboratories than behind the podium."[4] McClintock worked closely with graduate students, and followed their research work from year to year. Her appointment was later renewed for another four years.

Barbara McClintock had been a staff member of the genetics department of the Carnegie Institution at Cold Spring Harbor for twenty-five years. When she "retired" in 1967 at age sixty-five, the Carnegie Institution changed her title to Distinguished Service Member. She held this honorary title until her death twenty-five years later.

For her, retirement did not mean golf, trips, or relaxing. In fact, McClintock never really retired. She continued to teach and do new research, working daily in her laboratory, just as she always had. Research was still her top priority.

McClintock made a point of keeping up-to-date

with science. Advances in molecular biology were a special interest. She visited Cold Spring Harbor's library four times a week and knew exactly when each new scientific journal would arrive. Besides teaching, she attended seminars regularly and asked insightful questions.

McClintock used her own system to make notes on articles, letters, and research papers. As she read, she underlined in red, green, or blue with colored pencils and a ruler. In her round, up-and-down handwriting, she added side notes in the margins—*EXP* for a possible experiment, or *IMP* for important.

McClintock's insatiable curiosity continued to lead her in many directions. When she met with friends in her lab and at small dinner parties, she enjoyed discussing a wide range of topics—politics, roses, reading, extrasensory perception, world events, books, music, and more. Meteorology, the study of weather, remained a lifelong interest. She kept records of major storms on her recording barometer.

While McClintock remained a private person all her life, more and more people were becoming aware of who she was. In the small town of Cold Spring Harbor, across the waters from the laboratory campus, Barbara McClintock became recognized as "the corn lady." Residents respected her reserve and independence. Although they never made a fuss, many townspeople considered her a local treasure.

After her retirement, more honors came her way. The first was from the National Academy of Sciences (NAS), founded more than one hundred years earlier under President Abraham Lincoln. The academy honored Barbara McClintock in 1967 with its Kimber Genetics Award. On the Kimber award certificate, she was called "imaginative and daring," and her studies on the structure and workings of chromosomes "unmatched in the history of cytogenetics."[5]

The University of Missouri presented Barbara McClintock with an honorary degree in 1968, recognizing her as a former member of the faculty. No mention was made of McClintock's abrupt departure from Missouri in 1941. Times were changing. At some universities, women now had a better chance than before to become permanent faculty members.

Despite these honors, McClintock's research was not broadly recognized. She was still regarded as something of an outsider in the scientific community. With her uncanny thinking skills, it may have appeared to some people that ideas simply "came" to her.

Close friends and colleagues knew better. They had seen her extraordinary dedication to science. They knew her as an exacting person who spent long hours alone in the lab immersed in the smallest of details. McClintock had high powers of concentration. She drew connections with great insight. From college on, McClintock had relentlessly pursued new ideas about cells and genes. She was well equipped to build on past discoveries and forge innovative pathways to develop new theories.

"If she saw things the rest of us did not," said Nathaniel Comfort of the Center for History of Recent Science at George Washington University, "it was simply because she thought more clearly and deeply than most of us."[6]

Now there were hints that McClintock's "jumping genes" might be found elsewhere, not just in corn. Many molecular geneticists began taking a second look, some starting to make connections between her work and theirs.

As her next-door neighbors, Thomas R. Broker and his wife, Louise T. Chow, saw a steady pilgrimage to McClintock's lab. She hosted many people there for hours at a time. "If the conversation was good," said Thomas Broker, "it would cover a wide range of topics with emotion, intensity, and heartfelt concern about the issues."[7]

The trickle of recognition that began in the late sixties would become a flood. With typical reserve, Barbara McClintock would shun the spotlight.

10

Awash in Honors

Even in the late 1960s, many people still thought that McClintock's findings were true only for maize. The electron microscope, however, began yielding exciting evidence in the 1970s. Suddenly it was clear that certain segments of chromosomes in bacteria and other organisms could move about.

For close to three decades, McClintock had labored in obscurity, her work and its potential ignored. Now it seemed that McClintock had been right all along. The idea that "jumping genes" might be found in all living things became a real possibility.

A lesser person might have been tempted to say "I told you so." Not McClintock. As "jumping genes" were being found in other organisms, she was genuinely delighted.

McClintock's striking genetic discoveries began to be confirmed and her work was being more widely recognized. In 1970, she was honored with the National Medal of Science, the nation's highest award for outstanding science achievement. She was the first woman to receive this medal, viewed by many as the American version of the Nobel Prize.

She had earned this recognition, but she was simply not interested in fame. "She didn't see herself as a standard for women," said Susan Cooper, former director of public affairs at Cold Spring Harbor Laboratory. "She became a standard for women because of that."[1]

In the 1970s and 1980s many visitors came to see McClintock. She hosted them in her lab or invited them to walk with her. She was unswerving in her support of younger scientists. One was Nina Fedoroff, now at Pennsylvania State University. Fedoroff had given up her flute for science and had given up frogs for corn. The first crop of corn she planted in 1979 came from seeds given to her by Barbara McClintock. "I didn't read McClintock's writings twice," said Fedoroff. "I read them over and over and over again. It's probably the hardest thing I've ever done, becoming a maize geneticist."[2]

Just as McClintock offered patient one-on-one teaching, often spending hours a day with people, she never stopped learning from others. She kept up with the field of molecular biology, always intrigued by new ideas. McClintock "very faithfully went to all seminars at Cold Spring Harbor," noted her colleague Thomas Broker. "She was always there, listening and [offering] an insight or cogent remark . . . [she] quickly grasped the molecular approaches to her cytogenetics."[3]

Barbara McClintock's invitations to colleagues often opened the door to long discussions. Fellow scientist Bruce Alberts received a handwritten note in May of 1978:

Sunday morning

Bruce,

Please come to my laboratory
for

1. *Better coffee*
2. *Fresh orange juice*
3. *Oatmeal or eggs*
4. *Toast or English muffins (or both)*
5. *Marmalade or honey (or both)*
6. *Conversation*

Barbara[4]

At seventy-seven years of age, McClintock still lived in a small, spartan cottage across the highway from the Cold Spring Harbor campus. Each morning, said Broker, "people would be plowing down the hill in their cars on way to work. She'd stand by the side of the road for as long as five minutes, waiting for a clearing, then sprint across Highway 25A."[5] People were relieved when she moved in 1979 to Hooper House located on the campus itself, not far from her lab.

She stayed fit with walking and aerobics. One

In 1973, the building called the Animal House, where McClintock did much of her work at Cold Spring Harbor, was rededicated as the McClintock Laboratory. Dr. McClintock spoke at the ceremony.

morning, a fierce winter storm hit Long Island. Broker was surprised to find the walkway to Demerec Lab, where he and McClintock worked, already cleared. McClintock had arrived before the grounds crew. She had "tunneled through a mountain of snow and flipped it over her head," said Broker. "Her toughness and attention to physical technique belied her seemingly frail frame."[6]

Broker also recalled her habit of collecting black walnuts that had fallen from trees behind Carnegie Dormitory. For weeks, the nuts dried in wire baskets under her lab benches. Then she opened the shells with a pick and hammer. The black walnut brownies that came from her oven had "a more intense walnut flavor than can possibly be imagined," said Broker.[7]

Children were drawn to McClintock. Once a seven-year-old girl, the daughter of a researcher, saw Barbara McClintock strolling the grounds.

"What are you doing?" asked the girl.

"Picking walnuts. Would you like to join me?" said McClintock. After the two gathered walnuts together, the scientist invited the girl to see her lab.[8]

Doug Gensel, now a Nevada veterinarian, met Barbara McClintock when he was just seven. His mother, Susan Gensel Cooper, worked at the lab. Gensel saw McClintock often. She was skilled at pitching their conversations to a level that was right for his age. "When she respected you and you respected her, she talked about everything. I can remember not really knowing at first what she did," said Gensel. Then he began learning in school about the importance of her work. He remembers other encounters through the years—how McClintock enjoyed a taco dinner he cooked for her and how she wrote recommendations for him for college applications. "She was a tremendous individual," Gensel recalled.[9]

McClintock was open to new ideas. Not bound by conventional thinking, she once told how she got rid of a persistent wart. She rubbed it with a copper penny, then threw the coin over her shoulder into her cornfield and never looked back. She would tell her dentist not to worry about hurting her. By focusing her mind, she did not feel any pain.

The names of many geneticists can be found in most encyclopedias from the late 1970s. Mendel,

Morgan, Sturtevant, Beadle, Watson and Crick of DNA fame, Jacob, Monod, and others are listed. Barbara McClintock's name is usually missing. Few history books and science texts of that time refer to her.

Why was McClintock's later work overlooked for so long? Probably for a mix of reasons. Women scientists did not achieve recognition as easily as men. McClintock did her research alone and did not publish extensively. She stuck with her microscope and maize even after other scientists had turned their electron microscopes to bacteria and viruses. She attended few major professional meetings in her field, saying they made her so excited she could not sleep.[10] Even at Cold Spring Harbor, she gave few reports at conferences.

Her personality may have also played a part. Barbara McClintock was not one to draw attention to herself or her work. She preferred being by herself or with small groups. She had no pretensions, no need to impress others. She lived her science. It absorbed most of her waking hours. Intense and immensely intelligent, she had no time for show-offs or braggarts.

Whatever the barriers, now they were falling. Larger honors and awards, some with prize money, began to come her way. In 1981, McClintock was recognized by her fellow geneticists. She and Marcus M. Rhoades shared the Thomas Hunt Morgan Medal given by the Genetics Society of America, to recognize a lifetime of contributions to the field of genetics.

The John D. and Catherine T. MacArthur Foundation presented McClintock with its first

McClintock taught with passion and intensity, as shown in this photo taken during a 1981 plant course at Cold Spring Harbor.

MacArthur Prize Fellow Laureate Award in 1981. The MacArthurs' son Roderick MacArthur wrote, "Quite simply, your Laureate Award is for you to use for whatever purposes you choose."[11]

Such an award was perfectly suited to Barbara McClintock. She found recognition taxing, and she did not want to be pressured to achieve. No one ever needed to tell McClintock to be productive. As her own boss, she drove herself, working twelve to sixteen hours every day. In forty years at Cold Spring Harbor, she had probably worked well over twice as long as the average full-time worker.

In just two months in 1981, she won three high honors. Besides the MacArthur Prize ($60,000 a year, tax free, for life), she received the Wolf Prize in Medicine ($50,000) and the Albert Lasker Basic Medical Research Award ($15,000). For a person who cared so little for material things, it must have been startling. Possessions were a bother. She didn't want to own anything. "These things never have been important to me." she said.[12]

For Barbara McClintock, the money didn't matter. She would continue her work because she felt that people had not grasped what she had tried to say decades before.

As McClintock's fame grew, the publicity brought an unwelcome invasion of privacy, although she was always polite. She felt it was too much at once. She reminded people that she was seventy-nine. "At my age I should be allowed to do as I please and have my fun."[13] For McClintock, having her fun meant work.

Instead, she was news. Photographers and reporters clamored for pictures and information.

McClintock received still more honors in 1982. She called her various trips to accept them "going to have my shirt stuffed."[14] Among them were honorary degrees from Yale University and from Cambridge University in England. She politely turned down other honorary degrees.

Even while recognizing McClintock's wish to escape publicity, people started wondering if she might receive the Nobel Prize. *Newsweek* magazine noted that "no fewer than 33 recipients of the Lasker Award have gone on to win the Nobel Prize."[15]

"As people were finding transposable elements in people, yeast, corn, many things, I began to see references to Barbara's work," said James Bonner, her longtime colleague from Caltech days. "Then I made a bet with myself that she would get the [Nobel] prize."[16]

In 1983, Bonner won that bet.

11

Winning the Nobel Prize

Listening to the radio on October 10, 1983, Barbara McClintock heard her own name as the winner of the Nobel Prize in Physiology or Medicine. McClintock was the first woman to receive this prize unshared. Other women honored with it had shared the award.

In the entire history of the Nobel Prize, Barbara McClintock was just the third woman to win an unshared Nobel Prize in any science. Only Marie Curie, in 1911, and Dorothy C. Hodgkin, in 1964, held that same honor. Their awards were in chemistry.

The Nobel Prize was named after Alfred Nobel, a Swedish scientist and businessman. As the inventor of dynamite, he became very wealthy. When Nobel died, in 1896, he left money earmarked for establishing the Nobel Foundation to award prizes each year in physics, chemistry, physiology or medicine, literature, and peace. (In 1969, a sixth prize was added, in economics.)

The morning Barbara McClintock learned she was a Nobel winner, she set out on her daily walk. As one who preferred to do things quietly, she needed time to think about how to deal with the coming fanfare. Along the way, the seasoned observer saw trees and flowers that were like old friends. She thought about baking her famous walnut bread later that afternoon if reporters would spare her the time.

At the hastily called press conference that day at Cold Spring Harbor, McClintock spoke about her work. "I have been so interested in what I was doing and it's been such a deep pleasure," she said, "that I never thought of stopping, and I never had a time when I felt I didn't know what I wanted to do next."[1]

McClintock told of her great respect for the Carnegie Institution. Asked what working at Cold Spring Harbor was like, she replied:

> *Marvelous. . . . I don't think there could be a finer institution for allowing you to do what you want to do. Now, if I had been at some other place, I'm sure that I would have been fired for what I was doing, because nobody was accepting it. The Carnegie Institution never once told me that I shouldn't be doing it. They never once said I should publish when I wasn't publishing. . . . [They] just let me do what I wanted and supported me all the way along the line. I can't thank Carnegie enough.*[2]

In her official statement, McClintock said, "It might seem unfair to reward a person for having so much pleasure over the years, asking the maize plant to solve specific problems and then watching its responses."[3]

The days from October 10 to December 10, 1983, were anything but routine for Barbara McClintock. She had little time for research. Instead, she was photographed, interviewed, and interrupted. She talked with reporters, carefully shopped for a dress, and responded to thousands of well-wishers.

"All of us at the Carnegie Institution rejoice in your Nobel Prize so richly deserved," read a telegram from James Ebert, president of Carnegie.[4]

Swamped with mail, she found an important letter from the Nobel Committee buried in a pile. It gave details of her trip to Sweden. McClintock wrote back, apologizing for her late response: "I have never had a secretary. Consequently there is no one here sufficiently acquainted with my affairs to sort the unexpectedly large number of telegrams and letters."[5]

In December, Barbara McClintock flew to Sweden to attend the Nobel festivities. Her niece Marjorie Bhavnani and Bhavnani's husband and son went along. Before the day of the official awards presentation, all the award winners (called laureates) presented Nobel lectures about their research. McClintock talked about the mutations in the forty plants that she had started on the seedling bench in the greenhouse in the winter of 1944–1945. The accidental discovery of bizarre stripes on the maize leaves was the beginning of her

Soon after Barbara McClintock heard about the Nobel Prize, she attended the birthday party of a longtime Cold Spring Harbor resident. As a joke, she wore a Groucho Marx disguise so she would not be recognized.

understanding of how some genes could regulate others—"jumping genes."

Temperatures were below freezing on December 10, the day the Nobel ceremony was held. Guests and Nobel Committee members gathered in the Stockholm concert hall. The laureates took their seats on the podium before the large audience.

When Barbara McClintock's name was called, she rose and took the arm of King Carl Gustaf, then walked down the stairs with him to receive her Nobel Prize certificate and gold medal. McClintock enjoyed novel experiences and was "most pleased to be escorted by the king," recalled Thomas Broker.[6]

In most years, the Nobel Prize ceremony is a proper affair, solemn and dignified. The 1983 award for Physiology or Medicine caught the imagination of the audience. When the king bestowed the prize, the audience was swept up in the excitement. Everyone knew the honor due this tiny, sun-wrinkled eighty-one-year-old scientist. They clapped so hard, the floor shook.[7]

Back home, McClintock happily waded back into her work. Although she knew and understood the honor of the award, she said, "But you don't need the public recognition. . . . You just need the respect of your colleagues."[8]

Barbara McClintock may not have been aware of the extent of her colleagues' respect before she won the Nobel Prize. Scientists, however, were well aware of her fine mind and singular discoveries. Now she would also be applauded by ordinary people around the world.

12

Passing the Baton

Now in her eighties, Barbara McClintock felt that the worldwide recognition of her Nobel Prize was more like a burden than an honor. She begrudged the time away from her experiments. Dealing with repeated questions and trying to simplify her discoveries for the public was wearing. "I don't like publicity. I don't like my time taken up with *me* . . . at my age, it's quite a chore."[1]

She knew that receiving the Nobel long after her myth-shattering discoveries was better. Had it been awarded earlier, it would have interfered with her work and teaching.

Nothing, not even the Nobel Prize, could keep Barbara McClintock from her laboratory. She soon

Barbara McClintock proudly holds her Nobel medal and diploma. Behind her is a bust of Alfred Nobel.

returned to her familiar routine, working twelve or more hours each day, six days a week at Cold Spring Harbor.[2]

Her lab was home to her, reflecting her sense of thrift and order. Drawers held collections of pencil stubs, buttons, and other small items, sorted into open boxes. No pictures decorated the walls in her lab or office. Her bookshelves held many books on corn that showed her fascination with the Mayas, the Aztecs, and other cultures that depended on corn for food. Her large library included many other topics. During a discussion, she often pulled out just the right book to lend to a visitor.

McClintock had a large and important collection of vintage microscopes. She also appreciated good-quality modern equipment. Although she was frugal, if something struck her fancy, she might buy duplicates. She had several small, handheld vacuums, perfect for quick cleanups in the lab. With a toaster and hot plate, she often cooked meals in the lab.

In front of large windows looking out on a lawn stood her dissecting microscope and an empty Dundee Marmalade jar for pencils and brushes. A stack of empty Godiva chocolate boxes was evidence of her invitations to other researchers to join her in the lab for afternoon tea and chocolates.[3]

The honors continued. In 1984 and 1985 alone, she was recognized by nine prestigious U.S. and international organizations. McClintock now held honorary degrees from all the major universities in the United States. She had received every major

American science prize and others from science organizations around the world.

McClintock was also nominated to the National Women's Hall of Fame in Seneca Falls, New York, which celebrates the achievements of more than one hundred distinguished American women. McClintock would join the ranks of Jane Addams, Grace Hopper, "Mother" Jones, Helen Keller, Eleanor Roosevelt, Wilma Rudolph, and others.

McClintock was tired of attending ceremonies in her honor. In March 1986, Susan Cooper, who was Cold Spring Harbor Laboratory's director of public affairs, went to Seneca Falls in her place. "I rarely speak for Dr. McClintock," Cooper told the gathering at the National Women's Hall of Fame, "and when I do it is often to say 'no.' It gives me great pleasure on this occasion to say 'yes' and accept this honor for her."[4]

Cooper described McClintock's everyday life with no lab technicians, secretary, assistants, or housekeeper. She told how the few breaks in McClintock's day came when she took a colleague's call, talked with young scientists, or looked into related research projects. Cooper mentioned the two trips McClintock still made each year to South America for maize research.

The day before, Barbara McClintock had said to Susan Cooper, " . . . tell them I'm feeble and crotchety." Then the eighty-three-year-old laughed. "Tell them I am honored and delighted to be in such fine company."[5]

Cooper accepted the glass crystal award on McClintock's behalf. When Cooper returned to Cold Spring Harbor, McClintock, in turn, presented the crystal to Cooper, saying, "I can come over and visit it any time."[6] Today, the angles of the glass catch the sun, throwing bright rainbows on Cooper's office walls at the Trudeau Institute, where she now works.

As this remarkable woman's fame grew, more and more people became interested in meeting her. McClintock posted an index card sign by her apartment door: PRIVATE RESIDENCE. DO NOT ENTER. In time, the thumbtack grew rusty, but the sign did sometimes discourage strangers from dropping in.

Nathaniel Comfort noted, however: "Those who raised a feather to this wall found it easily toppled. Behind it was a woman of great warmth, relentless intellectual energy, and open mind. She was willing to talk to anyone who really had something to say; her only intolerances were of stupidity and zealotry."[7]

McClintock still loved her walks. The nearby shore, woods, and fields offered her a treasure trove of curiosities. Even if puzzling over something from that day's work, she noticed the tiniest details of the natural world. She took deep pleasure in the world around her.

"No plant was too humble to escape her attention," remembered Anna Maria Skalka, McClintock's former neighbor. While McClintock was content to walk alone, often a friend or fellow scientist joined her. With easy enthusiasm, McClintock nudged people into noticing things. "It was delightful to walk along with her. . . ," said Skalka. "In a way, she invited you to be a child again, with her."[8]

Over the years McClintock was like the Pied Piper, introducing fellow walkers to galls—growths on plants or trees—poison ivy plants, or beautiful patterns of jewelweed blossoms. To her, a red flower on a delicate head of Queen Anne's lace or one red tulip in a sea of yellow ones signaled mutations. McClintock recognized these as transposable genetic elements, the genetic signature of the plant.

Pat Craig, an editor at the Carnegie Institution, met McClintock in 1985 to go over a manuscript about the scientist. "I will remember that day as one of the most extraordinary of my life," said Craig. "She was extremely kind, plying me with all sort of treats—she had a wicked sweet tooth; jellybeans were among her favorites—but this belied an incredibly wary and steely person. And curious! Never have I met a person with more curiosity about life."

Craig calls the scientist "one of the most original thinkers I have ever encountered." After the visit, Craig sent McClintock a box of homemade chocolate chip cookies. "She thanked me for the 'nut and chocolate bearing cookies,'" said Craig. "Now who else would call them that?"[9]

Unlike some famous scientists, McClintock never became stuffy or proud. She remained modest and candid throughout her life. While she had no tolerance for shirkers or people with shallow ideas, she was open and warm with young people and students if their questions and ideas were genuine.

An artist created a visual play on McClintock's "jumping genes," painting the spiraling double helix of DNA with two pairs of denim jeans hopping between them. The painting, titled *Jumping Jeans*, must have appealed to McClintock's sense of fun. It hangs today in the library at Cold Spring Harbor Laboratory.

Interacting with people individually or in small groups was one of her joys. She came to relish the role of teacher. Said Susan Cooper, "She was wonderful with students. She loved to teach them and bring them along," spending hours in one-on-one teaching.[10] She was eager to pass the baton, systematically sharing her knowledge of genetics and often supplying fellow scientists and younger researchers with seeds from her collection. "From my lab next door, I was impressed at the line at her door," said Thomas Broker.[11]

McClintock's simple apartment reflected her orderliness. She ironed her sheets and kept them in plastic bags. Her shirts, many with rounded collars, were always crisp. Everything was tidy.

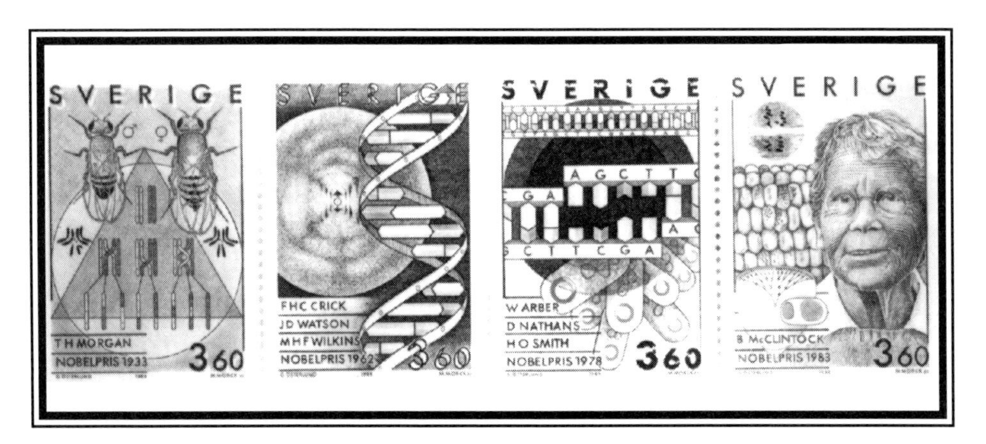

These 1989 Nobel stamps, issued in Sweden (Sverige), celebrate Nobel Prize work in genetics. "As far as portraits go, I share honors with the fruit fly," McClintock observed wryly.

McClintock loved the color red. Red canisters, dish rack and drain board, bowls, dishes, and bread box brightened her kitchen. Her sugar scoop and other utensils had bright red handles. McClintock's copper-bottomed cooking pans were top of the line, and she had a well-equipped electric range.

All her life, she remained captivated by tools and mechanical things. She had a fine set of wood saws. Well-designed gadgets intrigued her. When in-line skates came to Cold Spring Harbor, she asked her colleagues' children to skate back and forth so she could study precisely how the four-in-a-row wheels worked.

Friends convinced her to get a microwave oven. She was impressed as she learned to cook hot dogs, mushrooms, and baked potatoes in it. She never cooked corn, though. Ironically, this great maize geneticist had no taste for corn on the cob.

Barbara McClintock was well into her eighties before she got a telephone. When her family insisted that she have a phone, she finally gave in on the condition that no one would call her![12]

In 1987, a book of more than forty of her collected papers was published. At last, scientists and students had a single source for studying her challenging writings.

The Barbara McClintock Fellowship Fund was established in 1989 by the Carnegie Institution of Washington. The program honors McClintock and provides two-year fellowships for postdoctoral scientists at Carnegie. "The fellowships will continue as long as

Carnegie does," said Dr. Maxine Singer, president of Carnegie.[13]

Nearing her nineties, McClintock began to slow down physically. A heart problem sometimes made her pulse race. She learned how to slow it down by concentrating.

By the spring of 1992, Barbara McClintock could no longer take her customary walks around the Cold Spring Harbor campus. Her hearing and eyesight were failing, but her mind was as alert as ever.

More than thirty people gathered to celebrate Barbara McClintock's ninetieth birthday in June 1992. They presented her with an uncommon gift, a book called *The Dynamic Genome: Barbara McClintock's Ideas in the Century of Genetics*, edited by Nina Fedoroff and David Botstein. Like McClintock's life, the pages span the age of genetics.

The Dynamic Genome centers on this singular woman who, by her keen intellect and sheer persistence, spurred remarkable advances in genetics. For the now-frail scientist, the book was at once a towering and tender tribute filled with scientific discourse along with warm memories and endearing anecdotes.

• • • • •

Reflecting on her long years of research, Barbara McClintock once said, "I've had such a good time, I can't imagine having a better one. . . . I've had a very, very satisfying and interesting life."[14]

Barbara McClintock died on September 2, 1992, in a Huntington, New York, hospital after a brief illness.

Her life was commemorated at a moving memorial service at Cold Spring Harbor two months later.

"When she died, it was almost as if I lost a member of the family," said veterinarian Doug Gensel.[15] Others felt that same hollow sense of loss.

Barbara McClintock was one of a kind. The impression she left on the world of science would last. She had passed her own unique view of the beauty of genetics and the marvels of the natural world on to thousands of friends, colleagues, and students. To McClintock, a lifetime of plunging into the unfolding wonders of science had been a joy.

A Scientific and Personal Legacy

Barbara McClintock has been called one of the most important figures in twentieth-century science. Her insight into genetics earned her worldwide recognition.

Science moves forward so fast that exciting findings can become old news within five years. Science textbooks usually cover just ideas, not the people who made the discoveries. Neither of these statements is true in Barbara McClintock's case.

Today, reference books give good summaries of her life. Science textbooks cite McClintock by name and discuss how she moved science forward. "Genetics would not occupy its present high estate," said her colleague Marcus Rhoades, "were it not for her magnificent and pioneering contributions."[1]

Through difficult times and good times, Barbara McClintock believed in herself. As a person, she went her own way and never felt the need to conform. As a scientist, she trusted herself to learn. She knew her plants would teach her something if she followed the trail of new evidence.

Always farsighted, McClintock asked what her findings could mean for the future. She encouraged students to think ahead—what would be the next basic problem in their field? Even if they could not imagine the answer then, she urged them to keep such questions in their minds.[2]

McClintock cared about the future of science and those who would become scientists. "Young people have to be motivated to know what they're doing . . . we need to have people who know organisms can do fantastic things."[3]

Today, Barbara McClintock's "jumping genes" help scientists understand disease in humans and animals. "Jumping genes" may prove useful in breeding plants that resist disease, are tolerant of different climates, and offer specific nutrients. Her work may help reveal how plants and animals evolve to adapt to shock or change in the environment.

The American Philosophical Society in Philadelphia, Pennsylvania, which made McClintock a member in 1946, houses her papers. If all seventy feet of records were put into a single file drawer, it would be as long as two school buses. The collection includes many research notes, card files, photographs, and

The Smithsonian Institution in Washington, D.C., honored Barbara McClintock with a special exhibit from 1987 to 1988 at its National Museum of American History. The Smithsonian, known as "the nation's attic," now houses Dr. McClintock's microscopes, the nameplate from her laboratory door, a pair of her eyeglasses, and more.

the draft for her Nobel lecture. Letters reveal the long-term affection and support of friends.

In *The Dynamic Genome*, Anna Marie Skalka noted that the scientific world has had some extraordinary individuals, "like Barbara, whose imagination or special insight allows them to . . . open truly pioneering paths."[4]

Carnegie Institution president Maxine Singer observed that McClintock's work is central to modern genetics and biology: "Now there is a lively, international field of research on [transposable elements]. We have come to appreciate that they play an important role in the evolution of genomes."[5] They may, as McClintock believed, play a key role in evolution.

All her adult life, McClintock went about her daily work with intensity and persistence. She loved her work and found beauty in the story told through her microscopes. Science drew her in, filled her, and satisfied her eager mind and hungry curiosity. Friends relished her enthusiasm.

"The fact that she is gone," said colleague Howard Green at McClintock's memorial service, "makes me think of an extinct species or a miraculous creation that will never again be seen in the world."[6]

Barbara McClintock touched the lives of many people. With quiet brilliance, she changed the world of genetics, lighting the way for future generations of scientists.

Chronology

1902—Born June 16 in Hartford, Connecticut.

1908—Family moves to Brooklyn, New York.

1916—Enrolls at Erasmus Hall High School.

1919—Enrolls at Cornell University in Ithaca, New York.

1923—Graduates with a degree in botany, Cornell University.

1925—Earns master's degree, Cornell University.

1927—Earns doctoral degree, Cornell University.

1929–1931—Publishes nine scientific papers, including one identifying the ten maize chromosomes, and another, with Harriet Creighton, proving that pieces of genetic material switch places from one chromosome to another ("crossing over").

1931–1936—Receives fellowships from the National Research Council, the John Simon Guggenheim Memorial Foundation, and the Rockefeller Foundation. Continues her research at the University of Missouri, Caltech, and Cornell, and in Germany. Discovers ring chromosomes and nucleolar organizer region. Names the breakage-fusion-bridge cycle.

1936–1941—Works with Lewis Stadler, University of Missouri, on X-rayed chromosomes.

1939—Elected vice-president of the Genetics Society of America.

1941—Joins research staff of the Carnegie Institution at Cold Spring Harbor Laboratory, Long Island, New York.

1944—Identifies the seven chromosomes of pink bread mold, *Neurospora;* becomes the third woman to be elected to the National Academy of Sciences; notices highly significant changes in a crop of maize.

1944—Starts research on mobile genes and controlling elements.

1945–1951—Elected first woman president of the Genetics Society of America.

1946—Elected to the American Philosophical Society.

1947—Achievement Award, American Association of University Women, Educational Foundation.

1951—Presents paper at Cold Spring Harbor describing transposons, "jumping genes."

1953—Publishes a second paper on transposons; James Watson and Francis Crick announce the structure of DNA.

1956—Presents another paper at Cold Spring Harbor describing "jumping genes."

1959—Elected to the American Academy of Arts and Science.

1963–1969—Trains geneticists in Latin America for the Rockefeller Foundation's Agricultural Science Program.

1965—Appointed Andrew D. White Professor-at-Large, Cornell University.

1967—Kimber Genetics Award, National Academy of Sciences.

1970—First woman to receive the National Medal of Science.

1973—McClintock Laboratory dedicated at Cold Spring Harbor Laboratory.

1979—Honorary Doctor of Science, Harvard University.

1980—Salute from the Genetics Society of America.

1981—MacArthur Prize Fellow Laureate Award, Wolf Prize in Medicine, Albert Lasker Basic Medical Research Award; shares Thomas Hunt Morgan Medal, from Genetics Society of America, with Marcus M. Rhoades.

1982—Honorary Doctor of Science, Yale University and Cambridge University, England.

1983—Wins Nobel Prize in Physiology or Medicine.

1986—Elected to the National Women's Hall of Fame, Seneca Falls, New York.

1987—Publication of *The Discovery and Characterization of Transposable Elements: The Collected Papers of Barbara McClintock.*

1992—Publication of *The Dynamic Genome: Barbara McClintock's Ideas in the Century of Genetics* commemorates her ninetieth birthday; Barbara McClintock dies September 2 in Huntington, New York.

Chapter Notes

Chapter 1. In the Spotlight

1. All quotes in this chapter are from the transcript of Barbara McClintock's Cold Spring Harbor Laboratory press conference, October 10, 1983 (Cold Spring Harbor Laboratory Archives).

Chapter 2. An Independent Child

1. Evelyn Fox Keller, *A Feeling for the Organism: The Life and Work of Barbara McClintock* (New York: W. H. Freeman and Company, 1983), p. 18.

2. Sharon Bertsch McGrayne, *Nobel Prize Women in Science: Their Lives, Struggles, and Momentous Discoveries* (New York: Carol Publishing Group, 1996), p. 148.

3. Ibid., p. 147.

4. McGrayne, pp. 147–148.

5. Keller, p. 20.

6. Ibid.

7. Barbara Shiels, *Winners: Women and the Nobel Prize* (Minneapolis: Dillon Press, 1985), p. 199.

8. Keller, p. 24.

9. Shiels, p. 199.

10. Keller, p. 27.

11. Ibid., p. 24.

12. Ibid., p. 27.

13. Kathleen Teltsch, "Award-Winning Scientist Prizes Her Privacy," *The New York Times*, November 18, 1981, p. B3.

Chapter 3. Drawn to Science

1. Evelyn Fox Keller, *A Feeling for the Organism: The Life and Work of Barbara McClintock* (New York: W. H. Freeman and Company, 1983), p. 26.

2. Kathleen Teltsch, "Award-Winning Scientist Prizes Her Privacy," *The New York Times*, November 18, 1981, p. B3.

3. Keller, p. 27.

4. Ibid., p. 28.

5. Barbara Shiels, *Winners: Women and the Nobel Prize* (Minneapolis: Dillon Press, 1985), p. 200.

6. Keller, p. 33.

7. *Barbara McClintock, Pioneer of Modern Genetics* (Pleasantville, N.Y.: Sunburst Communications, Inc., 1990), IMG Educators videotape.

Chapter 4. Genetics—Monastery Garden to Cornfield

1. Helena Curtis and N. Sue Barnes, *Invitation to Biology*, 5th ed. (New York: Worth Publishers, 1994), p. 287.

Chapter 5. Two Momentous Discoveries

1. Harriet B. Creighton, "Recollections of Barbara McClintock's Cornell Years," in *The Dynamic Genome: Barbara McClintock's Ideas in the Century of Genetics*, ed. Nina Fedoroff and David Botstein (Plainview, N.Y.: Cold Spring Harbor Laboratory Press, 1992), p. 16.

2. Nina V. Fedoroff, "Two Women Geneticists," *The American Scholar*, vol. 65, no. 4, Autumn 1996, p. 590.

3. Roger Segelken, "Obituary: Barbara McClintock," *Cornell Chronicle*, Cornell University, September 10, 1992, p. 2.

4. Barbara McClintock, "Chromosome Morphology in *Zea Mays*," *Science*, vol. 69, June 14, 1929, p. 629.

5. Creighton, p. 13.

6. Mordecai L. Gabriel and Seymour Fogel. *Great Experiments in Biology* (Englewood Cliffs, N.J.: Prentice-Hall, 1955), p. 268.

Chapter 6. A Difficult Decade

1. Evelyn Fox Keller, *A Feeling for the Organism: The Life and Work of Barbara McClintock* (New York: W. H. Freeman and Company, 1983), p. 50.

2. *Barbara McClintock, Pioneer of Modern Genetics* (Pleasantville, N.Y.: Sunburst Communications, Inc., 1990), IMG Educators videotape.

3. Kathleen Teltsch, "Award-Winning Scientist Prizes Her Privacy," *The New York Times*, November 18, 1981, p. B3.

4. Winifred Veronda, "James Bonner Recalls Nobel Laureate Barbara McClintock: Caltech's First Woman Postdoc," *Caltech News*, February 1984, vol. 18, no. 1, p. 3.

5. Keller, p. 68.

6. Nina V. Federoff, "Barbara McClintock (June 16, 1902–September 2, 1992)," *Genetics*, vol. 136, January 1994, p. 4.

7. McGrayne, p. 161.

8. Ibid., p. 144.

9. Ibid.

Chapter 7. A Home at Cold Spring Harbor

1. Memorandum from Milislav Demerec, March 12, 1942 (Cold Spring Harbor Laboratory Archives).

2. Sharon Bertsch McGrayne, *Nobel Prize Women in Science: Their Lives, Struggles, and Momentous Discoveries* (New York: Carol Publishing Group, 1996), p. 163.

3. Letter from Barbara McClintock (at Stanford) to Milislav Demerec, November 3, 1944 (Cold Spring Harbor Laboratory Archives).

4. McGrayne, p. 165.

5. Barbara McClintock, "The Significance of Responses of the Genome to Challenge" (Nobel lecture, December 8, 1983) in *The Dynamic Genome: Barbara McClintock's Ideas in the Century of Genetics*, ed. Nina

Fedoroff and David Botstein (Plainview, N.Y.: Cold Spring Harbor Laboratory Press, 1992), p. 176.

6. *Barbara McClintock, Pioneer of Modern Genetics* (Pleasantville, N.Y.: Sunburst Communications, Inc., 1990), IMG Educators videotape.

7. Ibid.

8. Evelyn M. Witkin, "Cold Spring Harbor 1944-1955: A Minimemoir," in *The Dynamic Genome: Barbara McClintock's Ideas in the Century of Genetics*, ed. Nina Fedoroff and David Botstein (Plainview, N.Y.: Cold Spring Harbor Laboratory Press, 1992), p. 116.

9. Ibid.

10. "The Achievement Award," *Journal of the American Association of University Women*, Summer 1947, p. 243.

11. Ibid.

Chapter 8. Convincing the World

1. Sharon Bertsch McGrayne, *Nobel Prize Women in Science: Their Lives, Struggles, and Momentous Discoveries* (New York: Carol Publishing Group, 1996), p. 168.

2. *Barbara McClintock, Pioneer of Modern Genetics* (Pleasantville, N.Y.: Sunburst Communications, Inc., 1990), IMG Educators videotape.

3. Mel Green, "Annals of Mobile DNA Elements in Drosophila: The Impact and Influence of Barbara McClintock," in *The Dynamic Genome: Barbara McClintock's Ideas in the Century of Genetics*, ed. Nina Fedoroff and David Botstein, (Plainview, N.Y.: Cold Spring Harbor Laboratory Press, 1992), p. 119.

4. *Barbara McClintock, Pioneer of Modern Genetics* (Pleasantville, N.Y.: Sunburst Communications, Inc., 1990), IMG Educators videotape.

5. From remarks by Oliver E. Nelson, Jr., Ph.D., at Dr. McClintock's memorial service, November 1992 (Cold Spring Harbor Laboratory Archives).

6. *Barbara McClintock* videotape.

7. Ibid.

8. Helena Curtis and N. Sue Barnes, *Invitation to Biology*, 5th ed. (New York: Worth Publishers, 1994), p. 287.

9. McGrayne, p. 169.

10. From remarks by Waclaw Szybalski, Ph.D., at Dr. McClintock's memorial service, November 1992 (Cold Spring Harbor Laboratory Archives).

11. Ibid.

12. Evelyn Fox Keller, *A Feeling for the Organism: The Life and Work of Barbara McClintock* (New York: W. H. Freeman and Company, 1983), p. 174.

Chapter 9. In the Lab and on the Trail of Corn

1. Report sent by Barbara McClintock to Dr. Ackerman, August 24, 1967 (Cold Spring Harbor Laboratory Archives), p. 2.

2. Anna Marie Skalka, "A Tapestry of Transposition," in *The Dynamic Genome: Barbara McClintock's Ideas in the Century of Genetics*, ed. Nina Fedoroff and David Botstein (Plainview, N.Y.: Cold Spring Harbor Laboratory Press, 1992), p. 181.

3. Letter from Barbara McClintock to Dr. James A. Perkins, President, Cornell University, May 17, 1965 (American Philosophical Society Archives).

4. Roger Segelken, "Nobel Laureate Barbara McClintock Did Her Early Research at Cornell," *Cornell Chronicle*, vol. 14, no. 8, October 13, 1983, p. 1.

5. The Kimber Genetics Award certificate, National Academy of Sciences, April 24, 1967.

6. Nathaniel C. Comfort, "Barbara McClintock, 1992," Cold Spring Harbor Laboratory Annual Report, p. xi.

7. Telephone interview with Thomas R. Broker, Ph.D., August 14, 1997.

Chapter 10. Awash in Honors

1. Susan Cooper interview, October 16, 1996, Spring Harbor Laboratory.

2. Patricia Parratt Craig, *Jumping Genes: Barbara McClintock's Scientific Legacy*, Perspectives in Science, vol. 6 (Washington, D.C.: Carnegie Institution of Washington, 1994), p. 10.

3. Telephone interview with Thomas R. Broker, Ph.D., August 14, 1997.

4. Bruce M. Alberts, note from Barbara McClintock, in *The Dynamic Genome: Barbara McClintock's Ideas in the Century of Genetics*, ed. Nina Fedoroff and David Botstein (Plainview, N.Y.: Cold Spring Harbor Laboratory Press, 1992), p. 277.

5. Telephone interview with Thomas R. Broker, Ph.D., April 22, 1997.

6. Thomas R. Broker, Ph.D., at Barbara McClintock's memorial service, November 1992 (Cold Spring Harbor Laboratory Archives).

7. Ibid.

8. David Zinman, "A Secular Monastery for Pioneers of Science," part 2, Newsday, October 11, 1983, p. 4.

9. Telephone interview with Doug Gensel, D.V.M., March 10, 1997.

10. Interview with Ronald Phillips, Ph.D., Regents' Professor, University of Minnesota, and Chief Scientist, U.S. Department of Agriculture, January 11, 1997, San Diego.

11. Letter to Barbara McClintock from the John D. and Catherine T. MacArthur Foundation, November 13, 1981 (American Philosophical Society Archives), p. 1.

12. Kathleen Teltsch, "Award-Winning Scientist Prizes Her Privacy," *The New York Times*, November 18, 1981, p. B3.

13. Ibid.

14. Evelyn Witkin, Ph.D., at Barbara McClintock's memorial service, November 1992 (Cold Spring Harbor Laboratory Archives).

15. Matt Clark with Dan Shapiro, "The Kernel of Genetics," *Newsweek*, November 30, 1981, p. 74.

16. Winifred Veronda, "James Bonner Recalls Nobel Laureate Barbara McClintock: Caltech's First Woman Postdoc," *Caltech News*, February 1984, vol. 18, no. 1, p. 3.

Chapter 11. Winning the Nobel Prize

1. From transcript of Barbara McClintock's Cold Spring Harbor Laboratory press conference, October 10, 1983.

2. Ibid.

3. Barbara McClintock's official Nobel Statement issued at Cold Spring Harbor Laboratory, October 10, 1983 (Cold Spring Harbor Laboratory Archives).

4. Telegram from Carnegie president James Ebert to Barbara McClintock, October 10, 1983 (Cold Spring Harbor Laboratory Archives).

5. Letter from Barbara McClintock to Professor Jan Lindsten, M.D., Nobelkomittén, Karolinska Institut, Stockholm, Sweden, November 5, 1983.

6. Telephone interview with Thomas R. Broker, Ph.D., August 14, 1997.

7. Sharon Bertsch McGrayne, *Nobel Prize Women in Science: Their Lives, Struggles, and Momentous Discoveries* (New York: Carol Publishing Group, 1996), p. 173.

8. From transcript of Barbara McClintock's Cold Spring Harbor Laboratory press conference, October 10, 1983.

Chapter 12. Passing the Baton

1. *Barbara McClintock, Pioneer of Modern Genetics* (Pleasantville, N.Y.: Sunburst Communications, Inc., 1990), IMG Educators videotape.

2. Roger Segelken, "Obituary: Barbara McClintock," *Cornell Chronicle*, September 10, 1992, p. 2.

3. E-mail to the author from Patricia Gossel, Curator, Division of Science, Medicine, and Society, National

Museum of American History, Smithsonian Institution, Washington, D.C., January 3, 1997.

4. Susan Cooper, transcript of speech at the National Women's Hall of Fame, March 8, 1986 (Cold Spring Harbor Laboratory Archives).

5. Ibid.

6. Interview with Susan Cooper, Cold Spring Harbor Laboratory, October 16, 1996.

7. Nathaniel C. Comfort, "Barbara McClintock, 1992," Cold Spring Harbor Laboratory Annual Report, p. ix.

8. E-mail to the author from Anna Marie Skalka, December 3, 1996.

9. E-mail to the author from Pat Craig, editor, Carnegie Institution of Washington, July 25, 1996.

10. Susan Cooper interview.

11. Telephone interview with Thomas R. Broker, Ph.D., August 14, 1997.

12. E-mail to the author from Susan Cooper, January 4, 1998.

13. Press release, Carnegie Institution of Washington, June 12, 1990.

14. McGrayne, p. 173.

15. Telephone interview with Doug Gensel, D.V.M., March 10, 1997.

Chapter 13. A Scientific and Personal Legacy

1. Marcus Rhoades, "Barbara McClintock: Statement of Achievements, 1969" (Cold Spring Harbor Laboratory Archives).

2. *Barbara McClintock, Pioneer of Modern Genetics* (Pleasantville, N.Y.: Sunburst Communications, Inc., 1990), IMG Educators videotape.

3. Ibid.

4. Anna Marie Skalka, "A Tapestry of Transposition," in *The Dynamic Genome: Barbara McClintock's Ideas in the Century of Genetics*, ed. Nina Fedoroff and David Botstein

(Plainview, N.Y.: Cold Spring Harbor Laboratory Press, 1992), p. 181.

5. E-mail to author from Dr. Maxine Singer, March 16, 1997.

6. Howard Green, Ph.D., at Barbara McClintock's memorial service, November 1992 (Cold Spring Harbor Laboratory Archives).

Glossary

allele—One form of a gene. In Mendel's experiment (see page 32), *R* (round) and *r* (wrinkled) are alleles of the gene for seed coat texture.

bacteria—Tiny organisms that have no organized nucleus and multiply rapidly by splitting. Most are single cells.

botany—The scientific study of plants. Botanists study plants.

breakage-fusion-bridge cycle—A cycle where chromosomes break (by various means such as X-rays), the broken ends fuse, and then the chromosomes form a bridge when the fused chromosomes try to separate. The bridge then breaks, starting the cycle again. This cycle causes massive mutations in maize.

cell—The smallest unit of living things. Cells grow and multiply, carrying out the functions of plants or animals. All living things are either single-celled or combinations of cells.

chromosome—Tiny threadlike structures that carry genes. Chromosomes are found inside a cell's nucleus and hold information about the cell.

crossing over—During meiosis, the exchange (crossing over) of genetic material between paired sections of like chromosomes. Crossing over causes a genetic change in the next generation of offspring.

cytology—The study of the cell. Cytologists study cells' structure, life history, and functions.

DNA (deoxyribonucleic acid)—The chemical substance in cells that contains the information of life. Genes are made up of sections of DNA.

dominant trait—A trait that can mask the effect of a recessive trait.

gene—The unit of heredity. Genes are made up of DNA and are found on chromosomes.

genetics—The study of how traits are passed from one generation to the next in living things.

genome—A cell's complete genetic information—chromosomes and their genes.

heredity—The passing of traits from parents to offspring by means of genes on the chromosomes.

"jumping genes"—Nickname given by the press to transposons. *See* transposons.

maize—Indian corn.

meiosis—A special type of cell division, usually found in reproductive cells, that cuts the total number of chromosomes in an individual cell in half. Each cell produced through meiosis carries one *allele* of every gene the parent cell contained. When two cells that have undergone meiosis (such as a sperm and an egg cell) meet, they combine to form a cell with two *alleles* of each gene.

mutation—An accidental change in parent genes or chromosomes that can be inherited by the offspring. Events such as exposure to X-rays or certain chemicals can cause mutations.

nucleolar organizer region (NOR)—A small structure that must be present for a proper nucleolus to be formed.

nucleolus—A small body inside the cell's nucleus.

nucleus—A complex structure inside a cell where DNA is stored. Usually spherical, the nucleus controls the metabolism, growth, reproduction, and heredity of the cell.

recessive trait—A trait that is masked when combined with a dominant trait. Recessive traits are expressed only if an individual carries two copies of the recessive trait.

ring chromosome—A section of a chromosome whose frayed ends have joined to form a ring.

trait—A distinctive characteristic or quality.

transposon—A mobile element; a DNA piece that can move to a new site on a chromosome or to another chromosome. These "jumping genes" can change genetic instructions, causing changes in offspring.

Further Reading

Balkwill, Frances R. *Cells Are Us*. Minneapolis: Carolrhoda Books, Inc., 1993.

Dash, Joan. *The Triumph of Discovery: Women Scientists Who Won the Nobel Prize*. Englewood Cliffs, N.J.: Julian Messner, 1991.

Ganeri, Anita. *What's Inside Plants?* New York: Peter Bedrick Books, 1993.

Klare, Roger. *Gregor Mendel: Father of Genetics* (Great Minds of Science series). Springfield, N.J.: Enslow Publishers, Inc., 1997.

Shiels, Barbara. *Winners: Women and the Nobel Prize*. Minneapolis: Dillon Press, Inc., 1985.

Stidworthy, John. *Plants and Seeds* (Through the Microscope series). New York: Gloucester Press, 1990.

Tomb, Howard, and Dennis Kunkel. *MicroAliens: Dazzling Journeys with an Electron Microscope*. New York: Farrar, Straus and Giroux, 1993.

Wilcox, Frank H. *DNA: The Thread of Life*. Minneapolis: Lerner Publications Company, 1988.

On the Internet

Visit this site for lots of Barbara McClintock links:
<http://w3.aces.uiuc.edu/maize-coop/mcclintock.html>

Index

A
AAUW Achievement Award, 63–64
Albert Lasker Basic Medical Research Award, 88, 89
Alberts, Bruce, 83
alleles, 30
American Academy of Arts and Science, 72
American Association of University Women (AAUW), 63–64
American Philosophical Society, (APS), 63, 72, 106
Aztecs, 97

B
Barbara McClintock Fellowship Fund, 102
Beadle, George, 36, 49, 59, 86
Belling, John, 37–38
Bhavnani, Marjorie, 92
Blumenschein, Almiro, 74, 76
Bonner, James, 49, 89
botany, 27, 52
Botstein, David, 103
breakage-fusion-bridge cycle, 49, 54, 59
Broker, Thomas R., 80, 83, 84, 94, 101
Burnham, Charles, 36, 49

C
California Institute of Technology (Caltech), 48, 49, 68, 89
Cambridge University, 89
Carnegie Institution of Washington, 10, 56–58, 70, 77, 91, 100, 102
Central America, 72, 74
chromosomes, 9, 10, 28, 33, 35, 36–40, 43, 50, 61, 65
  knobs on, 39, 74, 76
  mutations, 48, 60, 100
  Neurospora (bread mold), 59, 60
  recombination, 41–45, 46
  ring chromosomes, 49
  X-rays, reaction to, 48, 54, 59
Cold Spring Harbor Laboratory, 7, 10, 55, 56, 57, 64, 66, 73–74, 78, 83, 86, 88, 91, 97, 98, 101, 102, 103
Comfort, Nathaniel, 79, 99
Connecticut, 13, 16
Cooper, Susan Gensel, 82, 85, 98, 99, 101
Cornell University, 24, 26–27, 35–36, 40, 45, 46, 48, 52, 55, 76–77
Craig, Pat, 100
Creighton, Harriet, 36, 40–45, 46, 68
Crick, Francis, 70, 86
"crossing over." *See* recombination.
cross-pollination, 43
Curie, Marie, 63, 90
cytology, cytologists, 27, 33

D
Darwin, Charles, 63
Demerec Lab, 84
Demerec, Milislav, 56, 57, 59
DNA, *deoxyribonucleic acid*, 63, 70, 86
Drosophila. See fruit fly.
The Dynamic Genome: Barbara McClintock's Ideas in the Century of Genetics, 103, 108

E
Ebert, James, 92
E. coli, 71
Einstein, Albert, 63
electron microscope, 59, 81
Emerson, Rollins, 50, 52
Erasmus Hall High School, 21–22

F
factor, 30
Fedoroff, Nina, 82, 103

Franklin, Benjamin, 63
fruit fly (*Drosophila*), 29, 33–34, 45, 49, 68

G

genes, 9–10, 27, 28, 30, 33–34, 36, 61, 63, 66, 68
genetics, 8, 12, 27–28, 33, 34, 36, 41, 43, 47, 105–106
  DNA, *deoxyribonucleic acid*, 63, 70, 86
  dominant genes, 30, 31
  genetic mapping, 61
  genome(s), 28, 108
  linked traits, 41
  recessive genes, 30, 31
  recombination, 1931 paper, 41–45
  transposons ("jumping genes"), 9–10, 64, 65–69, 73, 80, 81–82, 89, 94, 100, 106, 108
Genetics, 70
Genetics Society of America, 54, 62, 86
genome(s), 28, 108
Gensel, Douglas, 85, 104
Great Depression, 46, 52
Green, Howard, 108
Guggenheim Fellowship, 50
Gustaf, King Carl, 94

H

Handy, Benjamin, 12–13
heredity, 9, 28, 31, 35, 70
Hitler, Adolf, 50
Hodgkin, Dorothy C., 90
Holmes, Nigel, 66
honorary degrees and prizes. *See* McClintock, Barbara, honors.
Hooper House, 83

J

Jacob, François, 73–74, 86
John D. and Catherine T. MacArthur Foundation, 86
"jumping genes." See transposons.
Jumping Jeans, 100–101

K

Kato Yamakake, Takeo A., 74, 76
Kimber Genetics Award, 78–79
King, Carrie, 14
King, William, 14, 58

L

Lincoln, Abraham, 78
linked traits, 41, 43

M

MacArthur Prize Fellow Laureate Award, 88
MacArthur, Roderick, 88
maize, 8, 9, 35, 36–37, 39, 47, 60–62, 68, 76, 81, 92, 98
Massachusetts, 14
Mayas, 97
McClintock, Barbara
  Barbara McClintock Chair at Rutgers University, 62
  Barbara McClintock Fellowship Fund, 102
  birth, 13
  death and memorial service, 103–104
  early years, 13–14, 16–17, 19–20
  honors, 48, 50, 52, 54, 60, 62, 63–64, 72, 77, 78–79, 82, 86, 88, 89, 90–92, 94, 97–98, 103
  important papers, 39, 41–45, 68–69, 70, 71, 106, 108
  Nobel Prize, 7–11, 90–92, 94
  undergraduate years at Cornell, 24, 26–27
McClintock, Marjorie, 13, 22
McClintock, Mignon, 13, 22
McClintock, Sara Handy, 12, 13, 17, 22–23
McClintock, Thomas, 12, 13, 17, 22, 24
McClintock, Tom, 13
meiosis, 43
Mendel, Gregor Johann, 29–31, 33, 34, 36, 85
meteorology, 29, 78
microscopes,
  electron, 59, 81
  optical (light), 8, 38, 58
migration maps, 76
molecular biology, 59, 78
Monod, Jacques, 73, 74, 86
Morgan, Lillian, 34

Morgan, Thomas Hunt, 29, 33–34, 36, 45, 49, 50, 52, 86
mutations. *See* chromosomes.

N

National Academy of Sciences (NAS), 60, 72, 78
National Medal of Science, 82
National Research Council (NRC), 48, 50, 59
National Women's Hall of Fame, 98
Nelson, Oliver E., Jr., 68
Neurospora (bread mold), 59, 60
New York
  Brooklyn, 16, 21
  Ithaca, 24, 40
  Long Island, 55, 70, 84
  Manhattan, 16
Nobel, Alfred, 90
Nobel Prize in Physiology or Medicine, 7-11, 34, 63, 74, 82, 89, 90–92, 94, 95
  McClintock's life story for, 27
  Nobel Committee, 7, 8, 27
  Nobel Foundation, 91
  Nobel lectures, 92, 94
  Nobel statement, 92
nucleolus, nucleolar organizer (NOR), 50
nucleus, 28, 38, 50

P

Parker, Esther, 36
pea plants, 29–31
Pennsylvania, Philadelphia, 106
Perkins, James A., 77

R

recombination ("crossing over"), 1931 paper, 41–45
Rhoades, Marcus M., 36, 46–47, 55, 68, 86, 106
Rockefeller Foundation, 48, 52, 74
Rockefeller Foundation Agricultural Science Program, 74
Rutgers University, 62

S

Science, 39
Sharp, Lester, 27
Singer, Dr. Maxine, 103, 108
Skalka, Anna Marie, 76, 99, 108
smear (squash), 37, 38
South America, 72, 74, 98
Stadler, Lewis, 48, 50, 52, 54
Stanford University, 59
Stern, Curt, 45, 51
Sturtevant, Alfred H., 68, 86
Sweden, 92
Szybalski, Waclaw, 69–70, 71

T

Thomas Hunt Morgan Medal, 86
Time, 66
traits, 28, 30–31, 34, 41, 43
transposons ("jumping genes"), 9–10, 64, 65–69, 73, 80, 81–82, 89, 94, 100, 106, 108
  activator, 66
  dissociator, 66

U

University of Missouri at Columbia, 48, 53–55, 56, 57, 79

W

Watson, James, 29, 70, 86
Witkin, Evelyn, 61–62
Wolf Prize in Medicine, 88
World War II, 58

Y

Yale University, 89